Climate Change and the Media

Simon Cottle
General Editor

Vol. 27

The Global Crises and the Media series
is part of the Peter Lang Media and Communication list.
Every volume is peer reviewed and meets
the highest quality standards for content and production.

PETER LANG
New York • Bern • Berlin
Brussels • Vienna • Oxford • Warsaw

Climate Change and the Media

Volume 2

Edited by
Benedetta Brevini & Justin Lewis

PETER LANG
New York • Bern • Berlin
Brussels • Vienna • Oxford • Warsaw

Library of Congress Cataloging-in-Publication Data

Names: Brevini, Benedetta, editor. | Lewis, Justin, editor.
Title: Climate change and the media / edited by Benedetta Brevini
and Justin Lewis.
Description: Volume 2 | New York: Peter Lang Publishing, Inc., 2018.
Series: Global crises and the media; volume 27 | ISSN 1947-2587
Includes bibliographical references.
Identifiers: LCCN 2018005663 | ISBN 978-1-4331-5133-0 (pbk.: alk. paper)
ISBN 978-1-4331-5395-2 (hardback: alk. paper) | ISBN 978-1-4331-5436-2 (epdf)
ISBN 978-1-4331-5437-9 (epub) | ISBN 978-1-4331-6172-8 (mobi)
Subjects: Climatic changes—Public opinion.
Climatic changes—Press coverage.
Mass media and the environment.
Communication in the environmental sciences.
Classification: LCC QC903 .C557 2018 | DDC 070.4/4936373874—dc23
LC record available at https://lccn.loc.gov/2018005663
DOI 10.3726/b14826

Bibliographic information published by **Die Deutsche Nationalbibliothek**.
Die Deutsche Nationalbibliothek lists this publication in the "Deutsche
Nationalbibliografie"; detailed bibliographic data are available
on the Internet at http://dnb.d-nb.de/.

Table of Contents

Tables

Figures

Introduction

The first edition of *Climate Change and the Media* was published in 2009. That book was written during a period in which the global coverage of climate change had increased significantly (as James Painter and Mike Schafer's chapter in this volume documents). At the time, it appeared that in many countries, media coverage was, at last, *beginning* to reflect the scientific consensus about climate change (Zehr, 2009). So, while the 2009 book described the scale of the challenge—not least in a world where the abiding ideology of consumer capitalism is unhelpful in allowing us to understand and address the urgency of the problem—it contained more than a thread of optimism.

A few months after the book was published, the international climate talks in Copenhagen collapsed in disagreement, failing to set meaningful targets or legally binding limits on greenhouse gas emissions. Despite the increasing urgency of the timescale in to prevent significant climate change, hopes of a coherent, global vision to tackle the problem looked increasingly forlorn. Countries were left to continue business as usual, which in most cases meant taking little or no action to reduce emissions. For those scientists, politicians and NGOs who had brought us to the verge of an international agreement, this was a huge setback.

The other biggest news about climate change towards the end of that year was a story give the unoriginal (and inappropriate) sobriquet 'climategate', in which sceptics used the hacked emails of climate scientists at the University of East Anglia to claim that the climate change thesis was based on deception. Just when we thought that climate change sceptics were finally being consigned to their place on the wrong side of history, they were back dominating the terms of the debate. So when one of us was invited to take part in a discussion of the media coverage of climate change on the BBC in early 2010, the premise of the discussion—'Has the media coverage of climate change been too one-sided?'[1]—felt like a step backwards. Once again, those arguing that the news media should reflect the scientific consensus—and finally air a serious discussion about what to do reduce the scale and impact of climate change—were put firmly on the defensive.

Although subsequent inquiries into 'climategate' exonerated the scientists and found nothing to dislodge the careful weight of scientific evidence, the damage had been done. What might have been a period in which media coverage began to put pressure for action on climate change ended up mired in the same old controversy. Even now, as we enter the era of the Anthropocene, 'false balance'—where journalists give roughly equal weight to two sides regardless of the weight of evidence—continues to cast a shadow of inaction across the story of anthropogenic climate change.

This period was also one in which the economic crisis caused by a profligate banking system pushed governments around the world towards the pursuit of economic growth at any cost. In the grand narratives that dominate global news, the economic crash of 2008 and the subsequent squeeze on public finances meant that environmental concerns were cast aside—either ignored or attacked as a 'green tape' (meddlesome environmental regulations which held back business enterprise). Action on climate change was, once again, postponed—a delay that masks some difficult truths. The worst effects of climate change may be somewhere in the middle distance, but action to prevent it is needed well in advance of any tipping point. And in economies heavily dependent upon fossil fuels, the more economic growth we have, the more likely we are to increase greenhouse gas emissions, pushing any prospect of reducing them to sustainable levels even further beyond our reach (Jackson, 2010). As a consequence, after the initial furore of 'climategate', news coverage of climate change began to decline. Various studies have suggested that media coverage of climate change—and environmental issues more generally—peaked between 2007 and 2010—but has declined globally since then (Daly et al., 2015—a decrease documented in the UK by Vicky Dando in this volume). So, for example, Richard Thomas compared two full years of broadcast news in the UK (on the 10 pm weekday flagship bulletins on ITV and BBC) in 2007 and 2014. He tracked the coverage of over 30 topics and issues, and found that while the attention given to the economy increased significantly, environmental issues almost *disappeared* from news bulletins. Even during the high point of climate change coverage in 2007, the percentage of news time devoted to environmental issues was 2.5% on ITV and 1.6% on the BBC. By 2014, this had dropped to just 0.3% on the BBC and 0.2% on ITV (Lewis, 2015).

Decades of agenda-setting research show that public priorities are often more responsive to prominence in media coverage than they are to 'real world' trends (Lewis, 2001). The lack of media coverage means that while there is a degree of public concern about climate change (as Stuart Capstick and Lorraine Whitmarsh point out in this volume), it scarcely registers as a political issue in most countries. So, for example, it would not be unreasonable for people to assume that *the lack of coverage of climate change reflects a diminution of the threat*. We have thereby drifted into a cycle of relative silence, where lack of media coverage creates a sense of complacency in both public opinion and political debate.

Meanwhile, the dominance of consumerist discourses in commercial media—most of which are permeated with advertising messages encouraging us to carry on consuming—continues unabated. The ubiquity of advertising—on television, in news, magazines, the internet, social media and indeed, across our urban landscape—is generally regarded as an apolitical (if relentless) intrusion into our cultural environment. But the problem of climate change will not be *only* fixed by significant technological advances in energy production (Jackson, 2010)—it requires a turn away from the hyper consumption of wealthy nations. In this context, advertising acts as an ideological block on our imaginations, making a more sustainable future seem bleak (Lewis, 2013).

Moving forward to 2017, Post-truth was celebrated as word of the year by the Oxford and Macquarie dictionaries. Defined as 'relating to or denoting circumstances in which objective facts are less influential in shaping public opinion than appeals to emotion and personal belief' (Oxford Dictionaries, 2016) post-truth politics has thus been identified as a hallmark of the current era in the US and UK, while global politics seems to be threatened by the inescapable spread of fake news, misinformation, all amplified by social media echo chambers (Brevini & Woronov, 2017).

While on the one hand, post-truth politics is triumphing in contemporary political debates, climate change policy, from the moderate—if belated—success of the Cop 21 Paris Agreement, has moved into a new phase dominated by the withdrawal of the United States from an active role in progressive climate politics. In November 2017, at the U.N. climate summit in Bonn, Germany, the Trump administration made its official debut with a forum pushing coal, gas and nuclear power. The panel was entitled 'The Role of Cleaner and More Efficient Fossil Fuels and Nuclear Power in Climate Mitigation'; the forum was the only official act by the U.S. delegation during the UN climate summit.

The good news is that the panel was the only pro fossil fuel event of the summit, while new pledges were made by 19 countries, led by the UK and Canada within the new coalition *Powering Past Coal Alliance*, that includes Mexico, Angola, New Zealand, and Denmark.

Despite these pledges, the same governments that have signed the Paris Agreement on climate change—which aims to keep global warming below 2°C—continue to provide public funding to fossil fuel projects that could cause the world to exceed those climate targets. A report by a coalition of NGOs (Talk Is Cheap, 2017) found that the G20 countries provide four times more public financing to fossil fuels than to renewable energy. That means about $71.8 billion dollars of public finance for fossil-fuel projects per year between 2013 and 2015, compared with just $18.7 billion for renewable energy (Talk Is Cheap, 2017). In this context of political uncertainty, the role of the media is even more crucial.

One of the key issues examined by the new edition of Climate Change and the media is the way in which various media represent the issue of climate change to the general public. So, for example, to what extent is climate change still seen as a matter of scientific debate? How do external pressures—from corporate PR to NGOs—influence the debate? How prominent is the coverage of climate change in media coverage? To what extent is climate change presented as inevitable or, alternatively, which kinds of solutions receive the most coverage? And what impact does the coverage have on public understanding of the issue?

The book attempts to answer these questions from a range of different countries, allowing us to contrast the approaches taken by different media systems around the world. In particular, we can see the extent to which first world and third world media understand the problem differently, as well as explore differences between first world media (notably between the Anglophone countries and Europe).

The chapters embrace a range of methodologies, including content analyses of media coverage (news, internet, primetime television) from around the world, production analyses (examining the impact of the PR industry, the role of engineers, NGOs and attitudes of journalists in the UK and Canada) and audience reception. We will also examine a question that is generally overlooked: to what extent do the media industries—with their focus on built-in obsolescence and the ubiquity of media technologies—contribute to the global increase in carbon emissions?

In Chapter 1, Stuart Capstick and Lorraine Whitmarsh examine current research on public perceptions of climate change across several countries, analysing their findings to propose the most effective tactics for communicating climate change to audiences. They highlight research that shows how mitigation framings are more engaging for those with high levels of concern while adaptation framings are more engaging for less concerned individuals. It is not impossible to engage with sceptical audiences; using alternative framings, such as national security, patriotism or public health, can be effective in achieving support for climate change or renewables policy, or individual energy saving behaviours. Recent research also emphasises the role of visual imagery in generating—or impeding—public engagement.

In Chapter 2, Benedetta Brevini and Terry Woronov investigate how Australian politicians draw on propaganda strategies as they make claims about coal mining and the environment. Focusing on Adani Corporation's proposed Carmichael mine in Queensland, Brevini and Woronov highlight the oxymorons, lies, and relational strategies that demonise the opposition employed by political actors that aim at manufacturing an energy future that overlooks the climate crisis and celebrates coal. The Australian government's support for Adani's proposal is a prime example of political discourse on coal that posits a false equivalence between fossil and renewable-based fuels and eschews scientific, economic or environmental reality to instead generate a portrait of mining as essential for a proud twenty-first century Australia.

James Painter and Mike S. Schäfer compare international media coverage of climate change in Chapter 3, drawing out the differences between Anglophone countries, continental European countries, and the Global South, while examining the expanding role of rising digital-only players. Painter and Schafer show that, on the one hand, a number of important cross-national parallels exist in climate change coverage, where a 'societal turn' seems to have occurred in which coverage turned away from the scientific aspects of climate change towards the political and social ones. On the other hand, persistent differences remain between countries regarding the amount of space given to sceptical voices, the overall volume of climate change coverage in different countries, and the relative frequency of dominant frames used to interpret the topic other than in the coverage of set events or reports. New digital-only players, such as Huffington Post, Vice, and BuzzFeed, are also making inroads into the dominance of legacy media with a strong emphasis on coverage of the environment and a growing presence in a variety of languages and countries. But their relative strength and importance also varies between countries quite significantly.

Neil T. Gavin expands on this work on media coverage analysis by focusing, in Chapter 4, on the last COP21 Coverage in Paris by BBC websites. BBC web space was used to extend, to a very considerable degree, the amount of coverage the organisation was able to offer via the conventional terrestrial broadcasting route. While there were gaps in coverage, the web reports offered space for a great many institutions, organisations, individuals and political actors to have a say without an obvious or commanding preference for official and elite sources. A further contribution to public understanding was seen in the absence of climate sceptic voices.

Chapter 5 sees Anabela Carvalho and Eloisa Beling Loose highlight Brazil as an important case study for responses to climate change, given their responsibility for the Amazon as a carbon sink and their vulnerability to the consequences of changing climate. They explore studies of Brazil's coverage of the issue and link public perception to national policy decisions. The last few years have seen a series of setbacks in climate policies and other related measures in Brazil. However, the decline in political commitment stands in contrast with a positive 'climate of opinion' with strong public support for action. The Brazilian press also emerges as favourable to government-led mitigation policy, which compares positively with media in other countries.

Investigating the relationship between environmental journalism and politics in China in Chapter 6, Sibo Chen analyses the coverage of the 2015 United Nations Climate Change Conference. This chapter departs from liberal interpretations that consider the rise of environmental journalism in China as resistance against the country's authoritative media control. The overall aim is to elucidate environmental journalism's relationship to China's mounting ecological pressure and media political economy. For this purpose, the chapter analyses two Chinese media's coverage of the 2015 United Nations Climate Change Conference (COP 21): *People's Daily*, the national agenda-setting newspaper, and *China Daily*, the most-read English language newspaper published in China. By examining the ways in which the media texts depicted Chinese government's standpoints on climate change, the chapter shows how Chinese media present climate change to their audience by focusing on economic and political conflicts underlying climate negotiations. Such a discursive strategy further legitimizes China's aspiration for global environmental leadership.

Paolo Magagnoli assesses the 2013 documentary *Dirty Business* in Chapter 7, highlighting the nostalgic positioning of coal mining as 'the story of Australia' and the complete absence of climate change as typical of how apparently neutral media can narrow understanding of climate change politics. The story of *Dirty Business* exemplifies the ways in which historical knowledge shapes not only national consciousness but also climate change, politics, and alternative visions about the future. Far from galvanising transformative and revolutionary impulses, collective memory can also block positive utopian thinking and undermine the imagination of alternative and 'greener' economies and modes of production. Magagnoli suggests climate change advocacy should not rely solely on the collection and dissemination of more and better facts but instead look at the ways in which collective memory and national myths engage people at the level of emotion.

Lyn McGaurr and Libby Lester investigate the recent proliferation of 'last-chance travel lists' in Chapter 8, examining how these lists use imagery and language to convey information about climate change and highlighting the decay of the Great Barrier Reef as a case study. Last-chance travel lists spectacularise climate change by representing its impacts as imminent and dramatic, yet research suggests a strong focus on catastrophic consequences tends to make people feel impotent rather than empowered. Moreover, in their sample, dire linguistic predictions of disappearing global treasures were likely to be accompanied by commodified images of tourist destinations. While there is clearly a signal here about the importance and impact of climate change, the main call for action encourages tourists to use fossil fuels in an effort to consume them in haste.

Chapter 9 examines parallels between reporting on climate change and terrorism in Great Britain. Victoria Dando draws on analysis of over 1,000 newspaper articles between 1999 and 2012, suggesting that while narratives around terrorism normalise and amplify the threat, climate change narratives focus on future effects and fail to convey the same relevance and urgency, especially in tabloid and centre-right media coverage. Climate change is, like terrorism, a global issue, but unlike terrorism is shown to be something of a narrative nomad. If climate change is to learn from terrorism coverage, more can be done by climate communicators—and the media—to bring climate change closer to home. Making it nationally relevant, highlighting tangible impacts and actually emphasising risk may help build and sustain a compelling news rhetoric that more accurately reflects the scientific consensus and engages audiences on the issue.

In Chapter 10, Richard Maxwell and Toby Miller consider the media as polluting industries rather than sites for expressing or repressing the public interest about climate change. Since both paper-based and electronic media have been major polluters for a very long time, through energy consumption and emissions, materials, toxicity, and waste, climate change is not just a concept represented by the media, but a material reality, to which the media contribute as a dirty industry. To assess and mitigate the environmental impact of the media, Maxwell and Miller propose a form of very public green media accountancy that must include an element of public communication to emphasise the climatic impact of media work and its consumption and post-consumption.

In Chapter 11, Alana Mann interrogates the discursive framings of activists involved in resisting the development of coal seam gas in Australia and New Zealand, where attachment to 'place' plays a fundamental role, as the food and water security of local communities is threatened. The focus of the study is a grassroots social movement of over 250 groups in Australia and New Zealand, the Lock the Gate Alliance (LTGA). This movement is credited with uniting 'cockies, blockies, croppers and greenies' (Hutton, 2012) in non-cooperation campaigns against mining to protect water systems, agricultural land, bushlands, wetlands and wildlife, health, and Aboriginal and cultural heritage. Characterised by a network centric approach that includes tactics such as grassroots lobbying, physical protests and digital activism, the LTGA has been influential in winning support for its campaign to drive extractive industries off productive agricultural land.

Note

1. http://www.bbc.co.uk/programmes/b00qj1qs

References

Brevini, B. (2016). The value of environmental communication research. *International Communication Gazette, 78*(7), 684–687.

Brevini, B., & Woronov, T. (2017). Nothing but truthiness: Public discourses on the Adani Carmichael Mine in Australia. In B. Brevini & G. Murdock (Eds.), *Carbon capitalism and communication* (pp. 163–177). New York, NY: Palgrave Macmillan.

Carvalho, A. (2007). Ideological cultures and media discourses on scientific knowledge: Re-reading news on climate change. *Public Understanding of Science, 16*(1), 223–243.

Talk Is Cheap: How G20 Governments Are Financing Climate Disaster. (2017) Retrieved from http://priceofoil.org/content/uploads/2017/07/talk_is_cheap_G20_report_July2017.pdf

Daly, M., Gifford, L., Luedecke, G., McAllister, L., Nacu-Schmidt, A., Andrews, K., and Boykoff, M. (2015). World Newspaper Coverage of Climate Change or Global Warming, 2004–2015. Center for Science and Technology Policy Research, Cooperative Institute for Research in Environmental Sciences, University of Colorado, Web. [Date of access.] http://sciencepolicy.colorado.edu/media_coverage

Jackson, T. (2010) *Prosperity Without Growth? The transition to a sustainable economy.* London: Sustainable Development Commission

Lewis, J. (2001) *Constructing Public Opinion,* New York: Columbia University Press.

Lewis, J. (2013) *Beyond Consumer Capitalism: Media and the Limits to Imagination* London: Polity.

Lewis, J. (2015) 'Cycle of silence: the strange case of disappearing environmental issues' *The Conversation,* https://theconversation.com/cycle-of-silence-the-strange-case-of-disappearing-environmental-issues-36306

Zehr, S. (2009). An environmentalist/economic hybrid frame in US press coverage of climate change, 2000–2008 Boyce, T., & Lewis, J. (Eds.), in *Climate change and the media* (80–91). New York, NY: Peter Lang.

1. *Public Perceptions of Climate Change*

Stuart Capstick and Lorraine Whitmarsh

Introduction

Understanding public perceptions of climate change is critical in order to develop effective communication and educational strategies. Developing and implementing democratic policies and socially robust technologies to mitigate and adapt to climate change also requires an understanding of what the public think about climate change and how it should be tackled.

This chapter provides an overview of research on public perceptions of climate change, reviewing findings from quantitative (e.g., surveys, experiments) and qualitative (e.g., interviews) studies across several countries. We explore factors that shape perceptions, including the ways in which media coverage of climate change appears to be in informing and influencing public understanding of the issue.

We outline research on basic awareness and knowledge about climate change and then discuss research on people's attitudes and level of concern. We consider international research on these topics, with a particular focus on the UK and US. Research into the effects on public perceptions of climate change communication—via media and other sources—is then explored, before we conclude with implications of this research for communication efforts and future research.

Awareness and Knowledge of Climate Change

Surveys show that awareness and self-reported knowledge about climate change has been rising over the last three decades. In the UK, awareness of the terms 'climate change' or 'global warming' has been near universal since the early 2000s (around 99%; DEFRA, 2002, 2007; Whitmarsh, 2009). In the US, such basic awareness has also been present among a majority of the public since the early 1990s, though perhaps as many as 10% still had not heard or read anything about climate change by the mid-2000's (Nisbet & Myers, 2007). Worldwide, the degree to which publics have encountered the topic has been highly variable: whereas some awareness of climate change was widespread across developed countries by the mid-2000s, substantial proportions of people in developing countries still reported having encountered little or no information about the subject at this time (Brechin, 2010).

Although most members of the public do now accept the existence and human causation of climate change, there remains a substantial minority who are sceptical (e.g., Smith & Leiserowitz, 2012; Whitmarsh, 2011). In particular, despite overwhelming scientific consensus about the reality of climate change, under two-thirds of respondents in the US are of the view that climate change is even happening, with close to two-fifths also believing that scientists themselves are divided on this most basic of issues (Leiserowitz, Smith, & Marlon, 2010). Levels of climate change scepticism vary elsewhere in the world: cross-national work by Tranter and Booth (2015) for example, characterise only 4% of Germans and 9% of French respondents as 'sceptics', though some other countries—including the UK, Australia, and Norway—appear to hold levels of doubt similar to, or higher than, the US.

The precise meaning and use of the term climate change 'scepticism' with respect to public perceptions is not clear-cut (Capstick & Pidgeon, 2014; Hobson & Niemeyer, 2013) but one prominent framework distinguishes scepticism about rising temperatures ('trend scepticism') from doubts about the human causes ('attribution scepticism') and doubts relating to the severity and risks of climate change ('impact scepticism'; Rahmstorf, 2004); across these types, impact scepticism is found in the UK to be most commonplace, with trend scepticism being least prevalent (Poortinga, Spence, Whitmarsh, Capstick, & Pidgeon, 2011). Scepticism about climate change may partly be a product of the media presentation of climate change as controversial and uncertain (Boykoff & Boykoff, 2004; see also below), with this framing in turn exacerbated by the deliberate tactics of vested interests linked to the fossil fuel industry (Oreskes & Conway, 2010), and because the human causes of climate change are not self-evident. Yet, crucially, research highlights the interaction between values and scepticism: those with high pro-environmental values are less likely to be sceptical than those who value the environment less; and similarly, conservative political values are strongly associated with scepticism (e.g., Dunlap & McCright, 2008; Leviston & Walker, 2011; Whitmarsh, 2011). As discussed later, this highlights the tendency to interpret information about climate change in relation to one's existing views of the world.

Beyond a basic awareness of climate change, levels of knowledge and understanding are more variable. Most people in the UK, for example, claim they know 'a great deal' or 'a fair amount' about the issue in general (Whitmarsh, Seyfang, & O'Neill, 2011), although in respect of more specific terms (e.g., carbon emissions) or when not provided with a checklist of possible causes, levels of awareness are lower (DEFRA, 2007; Whitmarsh et al., 2011). Nevertheless, cross-country research carried out in 2014 concluded that public respondents were, in the main, reasonably well-informed about climate change, with a majority of people correctly answering questions about the causes and consequences of climate change (e.g., concerning the role of carbon dioxide and future increase in extreme weather events; Shi, Visschers, Siegrist, & Arvai, 2016).

Qualitative research indicates that the immediate associations that people hold with climate change are underpinned by recurrent themes; these concern changes in the weather and future climate impacts, the causal role of human activities and natural cycles, and the relationship between climate change and consumer society (Smith & Joffe, 2012; Tvinnereim & Fløttum, 2015). Other research has shown, similarly, that the public understands climate change as part of a broader set of social and environmental issues, such as air pollution, industrialisation, consumption and over-population rather than as a stand-alone issue (Darier & Schule, 1999; Tvinnereim, Liu, & Jamelske, 2017; Wolf & Moser, 2011). This conceptual linking reflects the media coverage which tends to discuss climate change in the context of local weather-related stories, such as UK flooding (Gavin, Leonard-Milsom, & Montgomery, 2011; Hargreaves et al., 2003) as well as people's underlying understanding of nature and natural systems (for example, as being cyclical or ever-changing; Connor & Higginbotham, 2013). People's conceptual associations are also likely to flow from the ways in which climate change, as a new concept, is made sense of in relation to familiar ideas and experiences (Ungar, 2000), a process known in social psychological literature as 'anchoring' (Breakwell, 1991; Whitmarsh et al., 2011).

Among the general public, there does appear to be a limited understanding of the relative contribution of different activities to climate change. In particular, there is an underestimation of the role of domestic energy use, meat eating/production, food miles and food waste (Attari, DeKay, Davidson, & De Bruin, 2010; Bailey, Froggatt, & Wellesley, 2014; DEFRA, 2007; Whitmarsh et al., 2011). Beyond this, there is also a general tendency to discount one's own contribution to causing climate change. People tend to identify causes of climate change with other people or countries, such as SUV drivers, corporations, the US or China (Lorenzoni & Pidgeon, 2006;

Whitmarsh, 2009), and with more 'distant' activities, namely industry and deforestation, rather than their own actions (Whitmarsh et al., 2011). This disconnection between individual actions and climate change is also reflected in British media coverage of the issue (Hargreaves et al., 2003), but as discussed later, the underestimation of personal energy use in contributing to climate change may reflect a strategy of reducing cognitive dissonance by deflecting responsibility for tackling climate change. The downplaying of individual-level emissions may also be underpinned by the sense that, in the context of a global and complex problem like climate change, any one person's actions can be viewed as negligible (Capstick, 2013).

Many have argued that the public will need to have a reasonable understanding of climate change and their role in causing and responding to it, particularly if certain policies (e.g., carbon taxation, the overhaul of energy systems) are realised or expanded. Indeed, research studies do appear to support the proposition that accurate knowledge of the causes of climate change underpins people's level of concern and support for policies to address it (Shi et al., 2016; Stoutenborough & Vedlitz, 2014). Nevertheless, a range of work indicates that there is a low salience of climate change overall (and, indeed, energy and environment) in individuals' day-to-day choices and actions (Upham et al., 2009; Whitmarsh, 2011). Growing awareness of environmental problems at an abstract or general level is often not translated into personally relevant decision-making or motivating attitudes—for example, whereas many 'green' consumers take small-scale personal action to reduce their environmental impact, they may do little or nothing to reduce more impactful behaviours such as flying or consumption of meat (Alcock et al., in press; Huddart Kennedy, Krahn, & Krogman, 2015; McDonald, Oates, Thyne, Timmis, & Carlile, 2015). When using energy, buying appliances or travelling, most do not consider carbon emissions (Upham et al., 2009). As we now discuss, climate change has limited relevance for many people in their everyday lives.

Attitudes to Climate Change

Beyond people's awareness and knowledge, research has also explored how the public evaluates climate change (i.e., their attitudes), including levels of perceived risk and concern. Overall, most people consider climate change to pose negative consequences (e.g., floods, drought; Lorenzoni, Leiserowitz, de Franca Doria, Poortinga, & Pidgeon, 2006; Poortinga & Pidgeon, 2003) but put in context, the issue is not a priority concern for most people (health, economic and social issues typically rank higher), and may not even be one of the public's main environmental concerns (Pidgeon, 2012).

The low ranking of climate change as a concern perhaps reflects a widespread perception amongst the public that the issue is a spatially and temporally remote risk, affecting future generations and other countries. Gifford and colleagues (2009) describe this as 'temporal pessimism' (environmental quality will decrease over time) and 'spatial optimism' (environmental quality worsens as geographic distance increases), which is evident in most countries and may stem in part from self-esteem associated with 'place identity' (Proshansky, 1978) as well as from media coverage of distant impacts. Whilst many perceive that climate change is already happening and will affect their local area (Spence, Poortinga, Butler, & Pidgeon, 2011), many people in developed countries do not feel it poses a prominent personal threat (e.g., O'Neill & Nicholson-Cole, 2009; Whitmarsh et al., 2011); although it may be seen as more proximal for some in developing countries, for example, for those reliant on agriculture (e.g., Basannagari & Kala, 2013; Le Dang, Li, Nuberg, & Bruwer, 2014). Climate change possesses many features that can lead to it being a psychologically 'distant' issue—that is to say, it is seen as something that occurs in remote locations, that will not affect people for some years, and that is uncertain or lacking relevance to one's own social group (Brügger, Dessai, Devine-Wright, Morton, & Pidgeon, 2015; Spence et al., 2011; Whitmarsh, 2015).

It is important to observe that much research shows that the public is highly heterogeneous in their attitudes to climate change—both within and between nations. Internationally, there is considerable variation in perceptions, with sub-Saharan Africa and South American countries showing relatively high (and rising) concern, compared to other parts of the world (Capstick et al., 2015). Furthermore, segmentation studies point to distinct attitudinal clusters, ranging from the most concerned and active to the disinterested or active deniers, indicating multiple 'publics' in respect of climate change engagement (Maibach, Leiserowitz, Roser-Renouf, & Mertz, 2011; Poortinga & Darnton, 2016). Belief in and concern about climate change varies according to a range of factors, such as gender, age, and ethnicity (e.g., Upham et al., 2011), although studies show values and ideology tend to be stronger predictors than demographic, knowledge, or other factors (Hornsey, Harris, Bain, & Fielding, in press; Whitmarsh, 2011). In particular, those with right-of-centre political beliefs and individualistic worldviews tend to be more sceptical, at least within the UK and US (Dunlap & McCright, 2008; Eurobarometer, 2009; Poortinga et al., 2011) with this effect appearing to have become more pronounced over time, at least in the US (McCright & Dunlap, 2011). In sum, in some countries in particular, there would appear to be an increasingly ideological component to climate change attitudes.

Communicating Climate Change

The way in which environmental change is communicated fundamentally shapes people's perceptions of the issue. Research also suggests that media coverage of climate change can influence individual-level behaviour, including intentions to reduce one's carbon footprint or involvement in environmental activism (Arlt, Hoppe, & Wolling, 2011; Feldman, Hart, Leiserowitz, Maibach, & Roser-Renouf, 2015).

For many issues, information sources, content, media, audience, and context interact to produce communication outcomes (Petty & Cacioppo, 1986). Since long-term global issues like climate change cannot be directly experienced, trust and credibility of second-hand sources become central influences on whether climate change is perceived as a genuine and serious risk. Surveys (Poortinga & Pidgeon, 2003; Whitmarsh, 2009) examining public trust in relation to particular sources of climate change information show that by far the most common sources of information on climate change are mass media—especially television and newspapers (Whitmarsh, 2009). However, trust in media sources is highly variable, whereas other sources (e.g., friends, family) may inspire more trust (Poortinga & Pidgeon, 2003).

As the most prevalent source of climate change information, the media is thus a salient influence on public perceptions (Hargreaves et al., 2003) and has played a major role in constructing climate change (e.g., Trumbo, 1996). Traditional media—including print and television—has drawn public attention to scientific and political debate—and often controversy—about climate change (e.g., Boykoff & Boykoff, 2004; Corbett & Durfee, 2004; Hargreaves et al., 2003). It has been argued, critically, that presentation of climate change (and other risk issues) as 'controversy' can strongly influence public perceptions about its reality and seriousness, and of the credibility of the scientific evidence and policy responses to it (Hargreaves et al., 2003). In particular, a concern with 'balance' when reporting climate science has portrayed the current state of knowledge to the public in disproportionately uncertain terms. Whereas there has for some time been a strong consensus among scientists that climate change is real, human-caused, and poses serious risks (e.g. Anderegg et al., 2010), this can be masked by media reporting that seeks to portray both 'for' and 'against' perspectives on the science (Boykoff, 2007; Boykoff & Boykoff, 2004; Boykoff & Mansfield, 2008).

The media's focus on climate change in the context of social and political developments (e.g., climate change legislation, extreme weather impacts on communities) and in terms of emotional representations (e.g., fear of future consequences, compassion towards the natural world) has, in parallel, cast the issue in tangible and human terms as well as in terms of an abstract science issue (Höijer, 2010; Peters & Heinrichs, 2004), using shared social symbols, experiences, and values to garner public interest (Hargreaves et al., 2003; McComas & Shanahan, 1999). Although this may enable a public audience to better apprehend the human dimensions and implications of climate change, it has also been suggested that emotionally loaded and fearful representations of risk can prompt a sense of disillusionment or powerlessness, particularly if there is no sense given of a recourse to meaningful action that can address it (Höijer, 2010; O'Neill and Nicholson-Cole, 2009; O'Neill et al., 2013).

Recent research highlights the key role of the media in shifting public concern away from environmental issues towards other (e.g., economic) ones. Following a peak in public environmental concern in 2007, numerous studies show public concern about climate change then dropped, while doubt and scepticism have risen, in many regions (Capstick et al., 2015b; Leiserowitz, Maibach, Roser-Renouf, Feinberg, & Howe, 2012a; Whitmarsh, 2011). While several explanations have been suggested for this widespread rise in doubt and apathy (see Pidgeon, 2012), most convincingly, Brulle, Carmichael, and Jenkins's (2012) analysis of U.S. opinion data from 2002 to 2010 show the economic recession, along with 'elite cues' (i.e., policy developments, politician statements), together shaped media coverage of climate change, which in turn influenced public attitudes towards climate change. Further research by Carmichael and Brulle (in press) also identified a pronounced effect of media coverage on public concern about climate change; this analysis likewise found that media attention to climate change was, in turn, a function of economic and political factors, as well as being influenced by the efforts of social movements and advocacy groups. Palm, Lewis, and Feng (2017) have likewise shown that the extent to which people attend to news and public affairs directly affects their views on climate change over time.

Despite clear indications of variability over time in levels of public concern about climate change, and the role of the media as an influence upon this, it is important to note that public audiences worldwide remain deeply concerned about environmental degradation as a result of human activities (e.g., Bulkeley, 2000; Fisher et al., 2012; Lorenzoni, Nicholson-Cole, & Whitmarsh, 2007). The influence of contextual factors on concern points to the problem of relying on public concern at only one point in time, as this may reflect only a superficial response to topical news events or recent experience. Furthermore, where people are given the opportunity to give their opinion in the context of an extended timeframe—as opposed to gauging people's most immediate concerns—the relevance of climate change as an issue is raised substantially (Capstick et al., 2015a).

For communicators and policy-makers interested in engaging the public with climate change, there is a growing body of evidence for how (and how not) to proceed—web-based resources that point to best practice in this area include www.connectingonclimate.org and www.climateoutreach.org. In respect of communicating climate change, the first clear message from research is that fear should be used with care, and perhaps avoided altogether (Feinberg & Willer, 2011; O'Neill & Nicholson-Cole, 2009) as—without a clear message about how to avoid the risks—it may be counter-productive. Framing information around positive messages of hope, rather than fear, seems to be far more likely to engage a range of audiences (e.g., Bain, Hornsey, Bongiorno, & Jeffries, 2011; Morton, Rabinovich, Marshall, & Bretschneider, 2011). Second, communicator characteristics are important. The credentials of the communicator are likely to be critical for message persuasiveness, with audiences believing messages from like-minded people, including politicians (Kahan, Jenkins-Smith, & Braman, 2010).

Third, communication should be targeted to those most likely to be receptive; and in trying to change attitudes, the most receptive audiences are those who are ambivalent or uncertain (Maio, Bell, & Esses, 1996), since existing attitudes and ideology act as a filter for new information (Kahan et al., 2010). Experimental work shows that, consistent with the well-established psychological phenomenon of 'biased assimilation' (Lord, Ross, & Lepper, 1979), people who are less sceptical about climate change evaluate the convincingness and reliability of newspaper editorials about climate change in a markedly different way to people who are more sceptical about climate change; in addition, these editorials tended to strengthen people's pre-existing views (i.e. to polarise opinion) (Corner, Whitmarsh, & Xenias, 2012). Given that the public is exposed to diverse, uncertain and conflicting claims about climate change through mass and new media (Boyce & Lewis, 2009), this research shows that media reporting on this topic risks reinforcing extant ideological divides over climate change. This has clear implications for communications intended to engage the public with climate change, as discussed below.

On the other hand, it is not impossible to engage with sceptical audiences; using alternative framings, such as national security, patriotism or public health, can be effective in achieving support for climate change or renewables policy, or individual energy saving behaviours (Feygina, Jost, & Goldsmith, 2010; Hoffman, 2011; Maibach et al. 2012). Bain et al. (2012), for example, found that framing policies to reduce greenhouse gases around impacts on personal relationships and economic/technological progress, shifted climate deniers' positions towards support for pro-environmental policies. Whitmarsh and Corner's (in press) UK experiment found that framing climate change in terms of either avoiding waste or supporting British industry elicited broad agreement and reduced scepticism amongst centre-right (i.e., more sceptical) participants, whereas a justice framing polarised audiences along political lines, appealing much more to those on the left-of-centre. Similarly, Howell and colleagues (2016) found that climate mitigation and adaptation framings affected participants' attitudes differently depending on their pre-existing levels of concern about climate change. Mitigation framings were more engaging for those with high levels of concern, whereas adaptation framings were more engaging for less concerned individuals.

Aside from seeking to emphasise specific aspects of climate change in order to promote engagement (the different ways of 'framing' climate change; Nisbet, 2009) researchers have also highlighted the importance of story-telling and narratives as a route to engage people with the topic in ways that scientific or fact-based information may fail to do (Lejano, Tavares-Reager, & Berkes, 2013; Markowitz, Hodge, & Harp, 2014). In a related way, recent research has emphasised the role of visual imagery in generating—or impeding—public engagement. For example, Chapman, Corner, Webster, and Markowitz (2016) highlight the importance of showing credible and authentic individuals when communicating climate change (including scientists as human subjects, as well as people experiencing or tackling climate change); by contrast, the use of 'climate clichés' (such as polar bears, melting ice, or power stations) was found to diminish the potential for public engagement.

Of course, there are limits to what can be done with information. Public engagement with low-carbon lifestyle change and broader societal change towards sustainability requires addressing the significant structural, social and financial barriers (Lorenzoni et al., 2007; Whitmarsh, O'Neill, & Lorenzoni, 2010). Also, critically, political distrust and disengagement need to be overcome, while acceptable and effective climate policies imply a greater degree of public participation in decision-making (Dietz & Stern, 2008).

Conclusion

This chapter has reviewed key findings from research on public perceptions and designing effective communications on climate change. The findings show widespread public awareness of climate change, though variable understandings which sometimes diverge with expert conceptions, and a strong ideological influence on attitudes. Information source, content, media, audience, and context interact to produce communication outcomes, with media acting as a strong—but not always trusted—source of information, and prior attitudes providing a strong filter for information processing. Experimental evidence shows that positive messaging, targeting receptive or ambivalent audiences, using like-minded or impartial sources, and employing narratives and authentic images, are amongst the ways to effectively produce engaging climate change communications.

References

Alcock, I., White, M., Taylor, T., et al. (2017). 'Green' on the ground but not in the air: Pro-environmental attitudes are related to household behaviours but not discretionary air travel. *Global Environmental Change.*

Arlt, D., Hoppe, I., & Wolling, J. (2011). Climate change and media usage: Effects on problem awareness and behavioural intentions. *International Communication Gazette, 73*(1–2), 45–63.

Attari, S., DeKay, M., Davidson, C., & De Bruin, W. (2010). Public perceptions of energy consumption and savings. *Proceedings of the National Academy of Sciences*, 107(37), 16054–16059.

Bailey, R., Froggatt, A., & Wellesley, L. (2014). *Livestock–climate change's forgotten sector: Global public opinion on meat and dairy consumption.* London, Uk Chatham House.

Bain, P. G., Hornsey, M. J., Bongiorno, R., & Jeffries, C. (2012). Promoting pro-environmental action in climate change deniers. *Nature Climate Change, 2*, 600–603.

Basannagari, B., & Kala, C. P. (2013). Climate change and apple farming in Indian Himalayas: A study of local perceptions and responses. *PLoS One, 8*(10), e77976.

Bostrom, A., Morgan, M. G., Fischhoff, B., & Read, D. (1994). What do people know about global climate change? 1. Mental models. *Risk Analysis, 14*, 959–970.

Boyce, T., & Lewis, J. (Eds.). (2009). *Climate change and the media.* New York, NY: Peter Lang Publishing.

Boykoff, M. T. (2007). From convergence to contention: United States mass media representations of anthropogenic climate change science. *Transactions of the Institute for British Geography, 32*(4), 477–489.

Boykoff, M. T., & Boykoff, J. M. (2004). Balance as bias: Global warming and the US prestige press. *Global Environmental Change, 14*, 125–136.

Boykoff, M. T., & Mansfield, M. (2008). 'Ye Olde Hot Aire': Reporting on human contributions to climate change in the UK tabloid press. *Environmental Research Letters, 3*(2), 1–8.

Breakwell, G. M. (1991). Social representations and social identity. *Papers on Social Representations, 2*, 198–217.

Brechin, S. R. (2010). Public opinion: A cross-national view. In C. Lever-Tracey (Ed.), *Routledge handbook of climate change and society* (pp. 179–209). London: Routledge.

Brügger, A., Dessai, S., Devine-Wright, P., Morton, T. A., & Pidgeon, N. (2015). Psychological responses to the proximity of climate change. *Nature Climate Change, 5*, 1031–1037.

Brulle, R. J., Carmichael, J., & Jenkins, J. C. (2012). Shifting public opinion on climate change: An empirical assessment of factors influencing concern over climate change in the U.S., 2002–2010. *Climatic Change, 114*, 169–188.

Bulkeley, H. (2000). Common knowledge? Public understanding of climate change in Newcastle, Australia. *Public Understanding of Science, 9*, 313–333.

Capstick, S. B. (2013). Public understanding of climate change as a social dilemma. *Sustainability, 5*(8), 3484–3501.

Capstick, S. B., Demski, C. C., Sposato, R. G., et al. (2015a). Public perception of climate change in Britain following the winter 2013/2014 flooding. In *Understanding Risk Research Group*. Cardiff University, UK.

Capstick, S. B., Whitmarsh, L., Poortinga, W., & Pidgeon, N. (2015b). International trends in public understanding of climate change over the past quarter century. *WIREs: Climate Change*. doi:10.1002/wcc.321

Carmichael, J. T., & Brulle, R. J. (2017). Elite cues, media coverage, and public concern: An integrated path analysis of public opinion on climate change, 2001–2013. *Environmental Politics.*

Chapman, D. A., Corner, A., Webster, R., & Markowitz, E. M. (2016). Climate visuals: A mixed methods investigation of public perceptions of climate images in three countries. *Global Environmental Change, 41*, 172–182.

Connor, L. H., & Higginbotham, N. (2013). 'Natural cycles' in lay understandings of climate change. *Global Environmental Change, 23*, 1852–1861.

Corbett, J. B., & Durfee, J. L. (2004). Testing public (un)certainty of science: Media representations of global warming. *Science Communication, 26*(2), 129–151.

Corner, A., Whitmarsh, L., & Xenias, D. (2012). Uncertainty, scepticism and attitudes towards climate change: Biased assimilation and attitude polarisation. *Climatic Change, 114*, 463–478.

Darier, E., & Schule, R. (1999). 'Think Globally, Act Locally?' Climate change and public participation in Manchester and Frankfurt. *Local Environment, 4*, 317–329.

DEFRA. (2002). *Survey of public attitudes to quality of life and to the environment: 2001.* London: Department for Environment, Food and Rural Affairs.

DEFRA. (2007). *Survey of public attitudes and behaviours toward the environment: 2007.* London: Department for Environment, Food and Rural Affairs.

Dietz, T., & Stern, P. C. (Eds.). (2008). *Public participation in environmental assessment and decision making.* Washington, DC: National Academies Press.

Dunlap, R., & McCright, A. M. A. (2008). Widening gap: Republican and Democratic views on climate change. *Environment, 50*, 26–35.

Eurobarometer. (2009). *Special Eurobarometer 300. Europeans' attitudes towards climate change.* Brussels: European Commission.

Feinberg, M., & Willer, R. (2011). Apocalypse soon? Dire messages reduce belief in global warming by contradicting just-world beliefs. *Psychological Science, 22*, 34–38.

Feldman, L., Hart, P. S., Leiserowitz, A., Maibach, E., & Roser-Renouf, C. (2015). Do hostile media perceptions lead to action? The role of hostile media perceptions, political efficacy, and ideology in predicting climate change activism. *Communication Research*. Advance online publication. doi:10.1177/0093650214565914

Feygina, I., Jost, J. T., & Goldsmith, R. E. (2010). System justification, the denial of global warming, and the possibility of 'system-sanctioned change'. *Personality and Social Psychology Bulletin, 36*(3), 326–338.

Fisher, A., Peters, V. Neebe, M., Vávra, J. Kriel, A., Lapka, M., & Megyesi, B. (2012). Climate change? No, wise resource use is the issue. *Environmental Policy & Governance, 22*, 161–176.

Gavin, N. T., Leonard-Milsom, L., & Montgomery, J. (2011). Climate change, flooding and the media in Britain. *Public Understanding of Science, 20*, 422–438.

Gifford, R. et al. (2009). Temporal pessimism and spatial optimism in environmental assessments: An 18-nation study. *Journal of Environmental Psychology, 29*, 1–12.

Hargreaves, I., Lewis, J., et al. (2003). *Towards a better map: Science, the public and the media*. London: Economic and Social Research Council.

Hobson, K., & Niemeyer, S. (2013). 'What sceptics believe': The effects of information and deliberation on climate change scepticism. *Public Understanding of Science, 22*(4), 396–412.

Hoffman, A. (2011). Talking past each other? Cultural framing of skeptical and convinced logics in the climate change debate. *Organization & Environment, 24*(1), 3–33.

Höijer, B. (2010). Emotional anchoring and objectification in the media reporting on climate change. *Public Understanding of Science, 19*(6), 717–731.

Hornsey, M. J., Harris, E. A., Bain, P. G., & Fielding, K. S. (in press). Meta-analyses of the determinants and outcomes of belief in climate change. *Nature Climate Change*. doi:10.1038/nclimate2943

Howell, R., Capstick, S., & Whitmarsh, L. (2016). Impacts of adaptation and responsibility framings on attitudes towards climate change mitigation. *Climatic Change, 136*(3), 445–461.

Huddart Kennedy, E., Krahn, H., & Krogman, N. T. (2015). Are we counting what counts? A closer look at environmental concern, pro-environmental behaviour, and carbon footprint. *Local Environment, 20*(2), 220–236.

Kahan, D. M., Jenkins-Smith, H., & Braman, D. (2010). Cultural cognition of scientific consensus. *Journal of Risk Research, 14*, 147–174

Le Dang, H., Li, E., Nuberg, I., & Bruwer, J. (2014). Farmers' perceived risks of climate change and influencing factors: A study in the Mekong delta, Vietnam. *Environmental Management, 54*(2), 331–345.

Leiserowitz, A., Maibach, E., Roser-Renouf, C., Feinberg, G., & Howe, P. (2012a). *Climate change in the American mind: Americans' beliefs and attitudes in September 2012*. Yale Project on Climate Change Communication, Yale University.

Leiserowitz, A., Smith, N., & Marlon, J. R. (2010). *Americans' knowledge of climate change*. Yale Project on Climate Change Communication, Yale University. Retrieved from http://environment.yale.edu/climate/files/ClimateChangeKnowledge2010.pdf

Lejano, R. P., Tavares-Reager, J., & Berkes, F. (2013). Climate and narrative: Environmental knowledge in everyday life. *Environmental Science & Policy, 31*, 61–70.

Leviston, Z., & Walker, I. (2011). *Baseline survey of Australian attitudes to climate change: Preliminary report.* Commonwealth Scientific and Industrial Research Organisation. Retrieved from https://publications.csiro.au/rpr/download?pid=csiro:EP105359&dsid=DS3

Lord, C. G., Ross, L., & Lepper, M. R. (1979). Biased assimilation and attitude polarization: The effects of prior theories on subsequently considered evidence. *Journal of Personality and Social Psychology, 37*(11), 2098–2109.

Lorenzoni, I., Leiserowitz, A., de Franca Doria, M., Poortinga, W., & Pidgeon, N. F. (2006). Cross-national comparisons of image associations with 'global warming' and 'climate change' among laypeople in the United States of America and Great Britain. *Journal of Risk Research, 9*(3), 265–281.

Lorenzoni, I., Nicholson-Cole, S., & Whitmarsh, L. (2007). Barriers perceived to engaging with climate change among the UK public and their policy implications. *Global Environmental Change, 17*, 445–459.

Lorenzoni, I., & Pidgeon, N. F. (2006). Public views on climate change: European and USA perspectives. *Climatic Change, 77*, 73–95.

Maibach, E. W, Leiserowitz, A., Roser-Renouf, C., & Mertz, C. K. (2011). Identifying like-minded audiences for global warming public engagement campaigns: An audience segmentation analysis and tool development. *PloS One, 6*, e17571.

Maibach, E. W., Nisbet, M., Baldwin, P., Akerlof, K., & Diao, G. (2010). Reframing climate change as a public health issue: An exploratory study of public reactions. *BMC Public Health, 10*, 1–11.

Maio, G. R., Bell, D. W., & Esses, V. M. (1996). Ambivalence and persuasion: The processing of messages about immigrant groups. *Journal of Experimental Social Psychology, 32*, 513–536.

Markowitz, E., Hodge, C., & Harp, G. (2014). *Connecting on climate: A guide to effective climate change communication.* New York and Washington, DC: Center for Research on Environmental Decisions and EcoAmerica. Retrieved from www.connectingonclimate.org

McComas, K., & Shanahan, J. (1999). Telling stories about global climate change: Measuring the impact of narratives on issue cycles. *Communication Research, 26*(1), 30–57.

McCright, A. M., & Dunlap, R. E. (2011). The politicization of climate change and polarization in the American public's views of global warming, 2001–2010. *Social Q, 52*, 155–194.

McDonald, S., Oates, C. J., Thyne, M., Timmis, A. J., & Carlile, C. (2015). Flying in the face of environmental concern: Why green consumers continue to fly. *Journal of Marketing Management, 31*(13–14), 1503–1528.

Morton, T. A., Rabinovich, A., Marshall, D., & Bretschneider, P. (2011). The future that may (or may not) come: How framing changes responses to uncertainty in climate change communications. Global Environmental Change, 21(1), 103–109.

Nisbet, M. C. (2009). Communicating climate change: Why frames matter for public engagement. *Environment: Science and Policy for Sustainable Development, 51(2),* 12–2

Nisbet, M. C., & Myers, T. (2007). The polls—trends: Twenty years of public opinion about global warming. *Public Opinion Quarterly, 71(3)*, 444–470.

O'Neill, S., & Nicholson-Cole, S. (2009). 'Fear won't do it': Promoting positive engagement with climate change through visual and iconic representations. *Science Communication, 30*, 355–379.

Oreskes, N., & Conway, E. M. (2010). Defeating the merchants of doubt. *Nature, 465*, 686–687.

Palm, R., Lewis, G. B., & Feng, B. (2017). What causes people to change their opinion about climate change? *Annals of the American Association of Geographers*. Advance online publication. doi:10.1080/24694452.2016.1270193

Peters, H. P., & Heinrichs, H. (2004). Expertise for the public: The science-journalism interface in German discourse on global climate change. *Public Communication of Science and Technology Conference (PCST-8)*, Barcelona, Spain.

Petty, R. E., & Cacioppo, J. T. (1986). *The elaboration likelihood model of persuasion*. New York, NY: Academic Press.

Pidgeon, N. (2012). Public understanding of, and attitudes to, climate change: UK and international perspectives and policy. *Climate Policy, 12*(S1), S85–S106.

Poortinga, W., & Darnton, A. (2016). Segmenting for sustainability: The development of a sustainability segmentation model from a Welsh sample. *Journal of Environmental Psychology, 45*, 221–232.

Poortinga, W., & Pidgeon, N. F. (2003). Public perceptions of risk, science and governance. Norwich: UEA/MORI.

Poortinga, W., Spence, A., Whitmarsh, L., Capstick, S., & Pidgeon, N. (2011). Uncertain climate: An investigation into public scepticism about anthropogenic climate change. *Global Environmental Change, 21*, 1015–1024.

Proshansky, H. M. (1978). The city and self-identity. *Journal of Environment and Behaviour, 10*, 57–83.

Rahmstorf, S. (2004). *The climate sceptics*. Potsdam Institute for Climate Impact Research, Potsdam. Retrieved from http://www.pik-potsdam.de/~stefan/Publications/Other/rahmstorf_climate_sceptics_2004.pdf

Safi, A. S., Smith, W. J., Jr., & Liu, Z. (2012). Rural Nevada and climate change: Vulnerability, beliefs, and risk perception. *Risk Analysis, 32*, 1041–1059.

Shi, J., Visschers, V. H. M., Siegrist, M., & Arvai, J. (2016). Knowledge as a driver of public perceptions about climate change reassessed. *Nature Climate Change, 6*(8), 759–762.

Smith, N., & Joffe, H. (2012). How the public engages with global warming: A social representations approach. *Public Understanding of Science, 22*, 16–32.

Smith, N., & Leiserowitz, A. A. (2012). The rise of global warming skepticism: Exploring affective image associations in the United States over time. *Risk Analysis, 32*(6), 1021–1032.

Spence, A., Poortinga, W., Butler, C., & Pidgeon, N. F. (2011). Perceptions of climate change and willingness to save energy related to flood experience. *Nature Climate Change, 1*, 46–49.

Stoutenborough, J. W., & Vedlitz, A. (2014). The effect of perceived and assessed knowledge of climate change on public policy concerns: An empirical comparison. *Environmental Science & Policy, 37*, 23–33.

Tranter, B., & Booth, K. (2015). Scepticism in a changing climate: A cross-national study. *Global Environmental Change, 33*, 154–164.

Trumbo, C. W. (1996). Constructing climate change: Claims and frames in US news coverage of an environmental issue. *Public Understanding of Science, 5*, 269–283.

Tvinnereim, E., & Fløttum, K. (2015). Explaining topic prevalence in answers to open-ended survey questions about climate change. *Nature Climate Change, 5*(8), 744–747.

Tvinnereim, E., Liu, X., & Jamelske, E. M. (2017. Public perceptions of air pollution and climate change: Different manifestations, similar causes, and concerns. *Climatic Change, 140*(3–4), 399–412.

Ungar, S. (2000). Knowledge, ignorance and the popular culture: Climate change versus the ozone hole. *Public Understanding of Science, 9*, 297–312.

Upham, P., Whitmarsh, L., Poortinga, W., Purdam, K., Darnton, A., McLachlan, C., & Devine-Wright, P. (2009). *Public attitudes to environmental change: A selective review of theory and practice.* A research synthesis for the Living with Environmental Change Programme, Research Councils UK. Retrieved from www.lwec.org.uk

Upham, P., Dendler, L., & Bleda, M. (2011). Carbon labelling of grocery products: public perceptions and potential emissions reductions. *Journal of Cleaner Production, 19*(4), 348–355.

Whitmarsh, L. (2008). Are flood victims more concerned about climate change than other people? The role of direct experience in risk perception and behavioural response. *Journal of Risk Research, 11*, 351–374.

Whitmarsh, L. (2009). What's in a name? Commonalities and differences in public understanding of 'climate change' and 'global warming'. *Public Understanding of Science, 18*, 401–420.

Whitmarsh, L. (2011). Scepticism and uncertainty about climate change: Dimensions, determinants and change over time. *Global Environmental Change, 21*, 690–700.

Whitmarsh, L. (2015). Analysing public perceptions, understanding and images of environmental change. In A. Hansen & R. Cox (Eds.), *Routledge handbook of environment and communication.* Abingdon, UK: Routledge.

Whitmarsh, L., & Corner, A. (in press). Tools for a new climate conversation: A mixed-methods study of language for public engagement across the political spectrum. *Global Environmental Change.*

Whitmarsh, L., O'Neill, S., & Lorenzoni, I. (Eds.). (2010). *Engaging the public with climate change: Behaviour change and communication.* London: Earthscan.

Whitmarsh, L., Seyfang, G., & O'Neill, S. (2011). Public engagement with carbon and climate change: To what extent is the public 'carbon capable'? *Global Environmental Change, 21*, 56–65.

Wolf, J., & Moser, S. C. (2011). Individual understandings, perceptions, and engagement with climate change: Insights from in-depth studies across the world. *Wiley Interdisciplinary Reviews: Climate Change, 2*(4), 547–569.

2. *Ignoring Climate Change, Celebrating Coal: Propaganda and Australian Debates on the Carmichael Mine*

Benedetta Brevini and Terry Woronov

2016 surpassed 2015 as the warmest year recorded since 1850, and 90% of the increase was due to the high levels of carbon dioxide in the atmosphere, levels not seen for 4 million years (Met Office, 2017). The consequences are far reaching, with sea levels rising and 'climate-related extremes such as heat waves, heavy precipitation and droughts increasing in frequency and intensity,' disrupting food production, accelerating species extinctions and the spread of diseases previously confined to the tropics (Brevini & Murdock, 2017; European Environmental Agency, 2017, p. 12). The latest COP 21 Paris agreement acknowledges that the CO_2 target will not keep the global temperature rise below 2°C, the level that was once set as the critical safe limit. Scientists concur that the planet will officially pass the 2°C level by 2036–2038 (COP21, 2015).

In spite of these frightening data, many industrial countries have refused to make serious commitments to move from carbon-based energy to renewable sources. Even more surprisingly, some nations are planning to increase their reliance on fossil fuels, as well as developing new sources of carbon-based energy. Australia is one such case. This chapter examines one particularly egregious example of Australia's ongoing commitment to fossil fuels: the Adani Corporation's proposed Carmichael mine in Queensland.

In late October 2010, the Indian energy behemoth Adani Mining Pty Ltd sought approval to build a massive greenfield coal mine in a remote area of north-central Queensland called the Galilee Basin. If built, the proposed

Carmichael Coal Mine would be the largest in Australia and one of the largest in the world, producing up to 60 million tonnes of coking (thermal) coal per year, targeted for export to India's coal-fired electricity generating plants. Since then, environmentalists have raised multiple concerns about the Carmichael mine and other proposed mining developments across the Galilee Basin. One is sheer size. The Carmichael mine would stretch approximately forty kilometres long by ten kilometres wide; over an estimated mine life of 60–90 years the mine could produce 2.3 billion tonnes of coal. Because the Galilee Basin is in a remote inland area, mined coal must be shipped to the coast for export, requiring the construction of a 189-kilometre rail line. Once at the coast, all the coal exported must be shipped through the Great Barrier Reef, requiring dredging of the reef to accommodate the larger container ships, as well as expanded port facilities. The Carmichael mine will also tap up to 12.5 billion litres per year of groundwater, drawn from Australia's Great Artesian Basin. This will lower the water table and influence agriculture and ranching across an enormous region of eastern Australia. Along with these immediate concerns, the Carmichael mine—indeed, the entire Galilee Basin—has been called a 'carbon bomb' for the enormous amount of CO_2 that burning the mined coal would release into the atmosphere. If burned, greenhouse gas emissions produced by the coal from this region could top 700 million tonnes per year, and 4.7 billion tonnes of emissions over its lifetime. 'This is more than 0.5% of the world carbon budget for limiting warming to 2°C … Over an average year, Carmichael's carbon emissions will excess those of Paris, New York, or Tokyo.' (Amos & Swan, 2015, p. 4; Foxwell-Norton & Lester, L. (2017).). A final cause for concern is that the Carmichael is only one of six proposed megamines for the Galilee Basin, where international mining consortiums have identified 247,000 square kilometres of coal. If the Carmichael is built, it will open the entire region for fossil fuel extraction.

Despite these well-documented environmental concerns (Taylor & Meinshausen, 2014), decreased demand for coal both domestically and internationally, and Australia's own ratification of the Paris Climate Accords, Australia has embarked on what some scholars have called a 'new coal rush.'[1] The production of coal is increasing and newly proposed megamines continue to be debated in state and federal parliaments, as well as in the public sphere.

Using a critical discourse analysis approach grounded on a critique of ideology, our research explores the propaganda strategies that are used to justify Australia's ongoing commitment to coal mining. Defining the purpose of propaganda, Bertrand Taithe and Tim Thorton (1999) argue that propaganda aims 'to convince, to win over and to convert; it has therefore to be convincing, viable and truthful within its own remit. … The shaping of the term propaganda is also an indication of the way the political nation judges the manner in which political messages are communicated. ... Propaganda promotes the ways of the community as well as defining them' (pp. 2–4).

As Jowett and O'Donnell (2005) powerfully argued in their book *Propaganda and Persuasion* (2005), a study of propaganda has to 'look for ideology in both verbal and visual representations that may reflect preexisting struggles and past situations, current frames of reference to value systems, and future goals and objectives. Resonance of symbols of the past encourages people to apply previously agreed-on ideas to the current and future goals of the propagandist' (Jowett & O'Donnell, 2005, p. 271).

In this chapter we expand on this work by seeking to understand how Australian politicians draw on propaganda strategies as they make claims about coal mining and the environment. We highlight the various discursive strategies employed by political actors that aim at manufacturing an energy future that overlooks the climate crisis and celebrates coal.

Methods

Drawing upon discourse generated in both the Australian Federal Parliament and the Queensland State Assembly,[2] as well as political reporting in the Australian media, we seek to understand how the nation's lawmakers develop a successful propaganda strategy that aims at reinforcing the future of coal. For the scope of the chapter, we adopt a definition of propaganda from Jowett and O'Donnell (2005): 'Our definition of propaganda focuses on the communication process—most specifically, on the purpose of the process: Propaganda is the deliberate, systematic attempt to shape perceptions, manipulate cognitions, and direct behavior to achieve a response that furthers the desired intent of the propagandist'" (Jowett & O'Donnell, 2005, p. 7).

We follow a brief discussion of our research methods with examples of political discourse that highlight three main propagandas strategies in the mining debates: relational strategies to demonize the opposition, oxymorons, and lies (Brevini & Woronov, 2017; Carvalho, 2005, 2007).

To understand how politicians use these strategies, we analysed political debates in Parliament around the Carmichael mine, as well as media coverage of these debates. We searched the Factiva and Lexis/Nexis CapitalMonitor databases for all references to the 'Carmichael Mine' and 'Adani' between January 1, 2010, the year the mine was proposed, and August 31, 2106. Because the Galilee Basin, the site of the proposed mine, is in Queensland, we searched both the Federal and Queensland Parliamentary debates, legislation, and law cases. References to the mine and the Indian conglomerate proposing the development were made by members representing all of Australia's major parties (Labor, Liberal, National, Greens), during a time period when government switched from Labor (2010–2013) to a Liberal-National Coalition (September 2013–present) under four different prime ministers.

Drawing on a variety of scholars from the tradition of critical discourse analysis (Carvalho, 2005, 2007; Fairclough, 1995; Van Dijk, 1993) the analytical framework employed specifically focuses on discursive strategies employed by political actors in Australia when debating the Adani mine (Carvalho, 2000, 2002). The concept of 'discourse strategy' in the context of propaganda is conceived as a form of manipulation of 'reality' by social actors that aim to achieve a certain goal. Critical discourse analysis, Van Dijk argues, aims to unveil the power of elitist groups or institutions that produce discourse that inevitably will 'enact, sustain, legitimate, condone or ignore social inequality and injustice' (1993, p. 252). Strategies of propaganda are thus inferred directly from texts authored by political actors and indirectly from quotes and reports of debates across four major newspapers: News Lts/Murdoch flagship outlet *The Australian*, *The Guardian* Australia edition, the *Sydney Morning Herald* and the major newspaper in Queensland, News Ltd's *Courier Mail.*

Relational Strategies to Demonize Opposition

In her investigation of climate change discourse in the UK, Carvalho (2005) suggests that *relational strategies* are frequently employed to frame climate change debates : 'Relational strategies are equivalent, but centred upon the speaker—how s/he constitutes her/himself into particular relations with others' (Carvalho, 2005, p. 9). In the case of the Carmichael mine, these strategies entailed using an ever-escalating set of negative terms to demonise opponents of the mine.

An early strategy revolved around re-framing the term 'activist' so that it connoted an enemy of both the Carmichael mine and the national interest. Examples abound in the media outlets we considered. On September 4, 2015, for example, Liberal Senator Matt Canavan told *The Australian* that 'Locals deserve more say than rich green elites,' warning that the green opposition will 'prevent the economic development of frontier parts of our nation, such as northern Australia' (*The Australian*, 2015a).

Another example from Assembly debate in Queensland shows how politicians construed green activists as enemies of development. This quote comes from debate over lawsuits filed by environmental groups to revoke the mining licenses granted to Adani:

> This has been a relentless, ideologically driven campaign. In fact, the 'lawfare' waged against Adani's Carmichael coalmine project has been one of the most coordinated, sophisticated and well-funded campaigns ever run by a collective of green activist groups against an economic development project in Queensland. However, what is certain is that that is a template on which we can expect future campaigns by the green activists to prevent future job creation opportunities to be based. (Cripps, 2016)

As lawsuits by e-NGOs continued to plague Adani and mine proponents in 2015–16, politicians' portrayal of environmentalists moved into the discourse of war and enemy combatants. Parliamentarians began to call members of green groups 'saboteurs' and 'terrorists,' and accused them of waging war on Australia's economy. Commonwealth Attorney General, Senator George Brandis, for example, stated: '[we must oppose the green's] declaration of war against the Australian economy' (August 20 2015). House of Representative MP George Christensen, a far-right member of the National Party, escalated the rhetoric, accusing environmentalists of being enemy combatants:

> [We must] end the ongoing warfare being waged by extreme green groups against major job-creating developments in the north. The recent setback with the Carmichael mine has alarmed many eager job seekers and business owners in the region who were looking at the opening up of the Galilee Basin and the expansion of the port of Abbot Point as a shot in the arm for our local economies. (19 August 2015)

By early 2016, environmentalists were not only extremists and enemy combatants, but were accused of terrorism: 'Their lies, misinformation, slander and the frivolous legal action attacking a company for the sake of furthering an ideological cause can only be described as terrorism if you look at the criminal code' (Christensen, House Hansard, February 8, 2016).

Australia, of course, is not the only place these relational strategies of demonising environmentalists has been used. Recent scholarly inquiry into accusations of 'eco-terrorism' (e.g., Potter, 2011; Sauvant, Fitzgibbon, & Nurse, 2016) however, has focused on whether or not environmental groups such as Earth First and various Animal Liberation groups are *really* terrorists (cites), concluding that the vast majority of environmental activism cannot be genuinely labelled as 'terrorism.' Indeed, the accusations of 'eco terrorism' and 'sabotage' directed at Carmichael mine opponents have no truth claims; this demonization of anti-mine activists is not linked to any actual forms of terrorist activities on the part of environmental groups.[3]

Yet arguing over the factual basis of the term 'eco-terrorism' misses the point. As Sauvant et al. (2016) note:

> In contemporary public discourse, the term 'terrorist' has strong negative connotations. By branding environmentalists as such and creating utilitarian legislation that *pre-emptively criminalises* (Fitzgibbon, 2004; italics added) and punishes for future acts, rather than a retributivist reactive approach to actual offending, governments aim to limit the extent of direct action by environmental activists. This manoeuvre helps support war on terror ideology and discourse, promoting the terror threat as real and imminent, therefore maintaining that all surveillance and other tactics are necessary to defeat the enemy. (2016, p. 79)

Oxymorons

A second propaganda strategy employed by political supporters of the Carmichael mine was the creation of new oxymorons in the debate on the mine. Benson and Kirsch (2010), in their investigation of the ways corporations use language to conceal or distract from possibly harmful activities, drew upon George Orwell to understand contemporary uses of euphemisms and oxymorons in political discourse. Orwell's classic *1984* described how the state sought to make alternative thought impossible through tactical appropriation and juxtaposition of key terms rather than outright censorship. One such appropriation was euphemisms, when the totalitarian state used terms that meant almost the exact opposite of what they appeared to mean. Another was the use of oxymorons, which are compound terms that combine contradictory messages. Benson and Kirsch (2010) point out that contemporary oxymorons are 'formulated and deployed to annul the implication that something may be harmful' (2010, p. 46). Their examples include 'Peacekeeper Missile,' a term from the Reagan era, and 'safe cigarettes' from tobacco companies today. These oxymorons rely on what they call 'figure/ground reversals' that call attention to one term while obscuring the other, thus 'conceal[ing] the contradictions of capitalism' implied (2010, p. 47).

Benson and Kirsch compiled articles on corporate oxymorons including: 'safe cigarettes' (Benson and Kirsch, 2010), and 'human energy' (Sawyer, 2010), in order to describe how corporations 'turn to a language of social responsibility to legitimise corporate activities with negative human and environmental consequences' (2010, p. 45). They call on scholars to pay attention to 'how the realm of images, languages and culture remain crucial to the operation and relatively unfettered expansion of multinational capital' (p. 46).

In the case of the Carmichael mine, several oxymorons show how mine proponents used oxymorons to conceal the harmful effects of the mine. Two were described in Kirsch's (2010) work: 'clean coal' and 'sustainable mining,' which global mining corporations use to 'neutralize critical discourse' (2010, p. 87). 'Clean coal,' promoted in the US 'as the solution to the energy crisis—even though it does not exist' (p. 88) was similarly deployed in Australia by proponents of the Carmichael mine, who frequently used this oxymoron to justify their support. One example was again from Attorney General George Brandis, who claimed: 'some forms of coal are cleaner than others and the coal that will be mined from the Carmichael mine by Adani is some of the cleanest coal in the world' (Brandis, Senate Hansard, November 23, 2015). Minister for the Environment Josh Frydenberg expanded on this argument:

> Australia's coal is predominantly low in sulphur and low in ash and can be used for these high-efficiency, low-emission power plants that we are seeing all around Asia which can reduce the carbon footprint by up to 40 per cent, and there are other new technology like carbon capture and storage which will be important to the future of clean coal. (Frydenberg, House Hansard, November 30, 2015)

This argument, repeated regularly in the halls of Parliament, was extended to claim that since Galilee Basic coal is the 'cleanest coal' in the world, and that once mined and burned it will actually *reduce* global warming. Minister for Northern Australia, Senator Matt Canavan, told the *Australian*: [If the Carmichael project doesn't go ahead] the coal will come from different countries without our environmental record, coal that is of lower quality … that will burn more carbon-dioxide emissions. This project is not just pro-development, it's not just anti-poverty, it's also pro-environment as well' (Geoff Chambers, *The Australian* April 8, 2016, Carmichael Coal Key to Easing Indian Poverty: Adani).

Countering the claims that Galilee Basin will produce 'clean coal,' research by The Australia Institute points out that it 'is only 10% above the average quality of domestic Indian thermal coal in terms of energy content' because 'the ash content of Carmichael coal is estimated to be 26%—more than double the average of 12% for Australian thermal coal' (Australia Institute, 2016, p. 5). The Institute report also pointed out that rather than helping the environment: 'Any environmental impact comparison would [also] need to account for the requirement that this low energy, high ash thermal coal needs to be transported 5–10 times the distance of domestic Indian coal—adding additional greenhouse pollution related to transport' (Australia Institute, 2016, p. 5).

'Sustainable mining' was a second, common oxymoron used by supporters of the Carmichael mine. As Kirsch (2010) describes, mining corporations globally have redefined the term 'sustainable' away from its original meaning, linked to environmentally sustainable projects, to now refer to economic variables, including job creation, royalties, and taxes. In Australia, mine proponents were quick to use this definition of 'sustainable mining' as well. In the words of Queensland Labor Minister of Natural Resources Anthony Lynham: 'If the state succeeds in opening up basins like the Galilee and the Surat on a sound, sustainable basis, thousands of jobs and economic prosperity will ensue. We are supporting the Galilee proponents to facilitate their projects on a commercial basis and almost all approvals are in place for the rail corridor' (Lynham Queensland Assembly May 7 2016).

Beyond drawing upon existing corporate oxymorons for their political arguments, Australian politicians also invented their own oxymorons to elicit support for the Carmichael mine (Woronov, 2017). As mine opponents filed multiple lawsuits in Commonwealth and Queensland courts to opposed mining licenses and environmental approvals for the Carmichael mine, supporters of the mine coined the oxymorons 'legal sabotage' and 'vigilante litigation' as part of their effort to de-legitimise legal challenges and demonize mine opponents. Attorney General George Brandis was a primary proponent of this new language, claiming that anti-mine lawsuits filed under the auspices of Australia's Environmental Protection and Biodiversity Act (1999) were a form of 'legal sabotage': '[The EPBC Act] has been availed of in recent years by vigilante litigants who are quite blatant, quite shameless, in their declaration that they intend to use that enhanced statutory standing for the purpose of economic destruction, and we will not allow that to occur' (Brandis September 8, 2015, Senate Hansard). Then-Prime Minister Tony Abbott reported to the Senate on August 19, 2015, 'This mine is being legally sabotaged by green activists waging a campaign against the coal industry.'

Lies

Parliamentarians and supporters of the Carmichael mine have also relied on mendacious statements to convince voters of their argument. Australia, once again, is by no means the only place this is occurring; as radical writer Henry Giroux and Bhattacharya note, 'lying is part of an anti-politics of performance and showmanship' (2017, p. 512) that turns lies into political spectacles.

> The best examples related to the Carmichael mine are the 'zombie facts' of the mine's benefits to Queensland and Australia. Most common are references to the number of jobs the Carmichael mine will provide to the Queensland economy, where the jobs situation is portrayed as desperate: 'The Adani Carmichael coalmine offers up to 10,000 new jobs, mainly in Queensland; $20 billion of investment in Australia; and power, to build the living standards of 100 million people in India.' (Landry, 2015)

This figure of 10,000 new jobs has been repeated by virtually every mine supporter since 2014, including the Prime Minister the Attorney General, and Liberal and National Party members of Commonwealth and Queensland Parliaments.

Once built, however, the mine is not projected to offer anywhere near 10,000 jobs. The Adani Corporation's own economist predicted that the total number is closer to 1,464: 'Adani's economist, Jerome Fahrer from ACIL Allen, found that Adani's mine and rail operations would employ around 1,800 people directly and create around 1,000 downstream jobs in "other services"' (Amos & Swann, 2015, p. 1). Moreover, in testimony before a Senate Committee on how the Queensland government spends monies from the federal government, it was reported that mining is a small fraction of the Queensland economy, and is only viable due to significant subsidies from the state. Economist Roderick Campbell declared:

> The coal industry actually only accounts for 1.2 per cent of employment in Queensland. It contributes around four per cent of Queensland government revenue. If we stop and think about that for a second, that means that basically 99 per cent of Queenslanders do not work in the coal industry …, Last year, the Australia Institute released research showing that over six years the Queensland state government had put around $8 billion into spending that was primarily aimed at benefiting the coal industry and we estimate that around $2 billion of that relates to developments at Abbot Point. … Deputy Premier Seeney said that the government was prepared to put in whatever it takes to make those projects viable. (Campbell, 2015)

In spite of these facts, the Minerals Council of Australia, the peak body representing the mining and minerals industries nationally, argued in support of the Carmichael mine in 2015: 'Mining's environmental footprint is tiny, and the benefits significant' (2015, p. 2).

Another example of the deployment of 'lies' are the ways politicians have tried to deflect the effects the Carmichael mine will have on the Great Barrier Reef. Although the Galilee Basin is inland from the coast, the mine will have devastating effects on the reef, which is already under threat from climate change, acidification, and agricultural run-off (Slezak, 2016). Shipping the mined coal to India will require a massive expansion of the coastal coal port at Abott Point, including dredging of parts of the reef, and the amount of coal shipped through the treacherous waters of the Great Barrier Reef will drastically increase if the mine is approved. More generally, the mine's contributions to global CO_2 emissions will contribute to rising temperatures that are already killing the reef.

Politicians, however, kept avoiding any discussion on climate change, thus spinning inconvenient facts by arguing that the mine's inland location absolves the Carmichael from any effects on the Great Barrier Reef. As Queensland Senator MacDonald explained:

> The Adani mine is some 300 kilometres from the coast. … Of course, 300 kilometres inland is nowhere near the Great Barrier Reef and all the protections in place to protect the Great Barrier Reef. (MacDonald, 2016)

Conclusion

In a theatrical gesture in February 2016, the Commonwealth Treasurer, Scott Morrison, brandished a lump of coal during question time in the Federal House of Representatives declaring: 'This is coal. Don't be afraid, don't be scared' (Murphy, 2017). He elaborated on his position in a radio interview a few days later, arguing that Australians should 'have no more fear of coal than we have a fear of wind, or solar, or wave energy, or pump-hydro, or whatever the option is' (quoted in Hunt, 2017). This statement, which positions coal as simply another option for energy generation alongside renewables, is a prime example of political discourse on coal that posits a false equivalence between fossil and renewable-based fuels, and ignores the overwhelming scientific evidence on the causes and consequences of climate crisis. Constant repetition of the highlighted propaganda strategies by political actors in Australia is perpetuating a 'dominant order of discourse' (Foucault, 1981,) where climate change is excluded from the debate on the energy future of the country.

Foucault warned that every body politic has 'its "general politics" of truth: that is, the types of discourse which it accepts and makes function as true; the mechanisms and instances which enable one to distinguish true and false statements' (Foucault, 1984, pp. 72–73). As we have shown, propaganda is always articulated around of system of truths and expresses a logic of exclusive representation. It is the purpose of propaganda to convince, to win over and to convert the public by employing a complex, overlapping set of discursive strategies that work together to produce very particular political effects. In this case, the political effect sought by politicians is support to build a greenfield megamine that will have devastating environmental impacts.

Current discourse on mining in the country eschews scientific, economic or environmental reality, to instead generate a portrait of mining as essential for the nation (Pearse, 2009) and mine opponents as elite, unpatriotic, uncaring, unmodern, evil.

Notes

1. See www.coalrush.net.
2. The state of Queensland has a unicameral legislative chamber called 'The Legislative Assembly."

3. In 2017 several environmental groups began "non-violent direct action" campaigns against Adani and other companies affiliated with the Carmichael mine. Many activists have been arrested for trespass and for impeding the normal course of business, but none have been charged with any violent offenses.

References

Amos, C., & Swann, T. (2015). Carmichael in context, quantifying Australia's threat to climate action. *The Australian Institute*. Retrieved from http://www.tai.org.au/sites/defualt/files/Amos%202015%20Carmichael%20in%20context%20-.pdf

The Australia Institute. (2016). The coal wars: Facts check. Retrieved from http://www.tai.org.au/sites/defualt/files/P303%20Coal%20hard%20facts_0.pdf

Benson, P., & Kirsch, S. (2010). Corporate oxymorons. *Dialectical Anthropology, 34*(1), 45–48.

Blewett, R. (2012). *Shaping a nation: A geology of Australia*. Canberra, Australia. Geoscience Australia and ANU E-Press.

Brevini, B., & Murdock, G. (2017). *Carbon capitalism and communication. Carbon Capitalism and Communication: Confronting Climate Crisis*, London: Palgrave Macmillan.

Brevini, B., Woronov, T. (2017). Nothing but Truthiness: Public Discourses on the Adani Carmichael Mine in Australia. In Brevini, B. and Murdock, G. (Eds.), *Carbon Capitalism and Communication: Confronting Climate Crisis*, London: Palgrave Macmillan. (147–159).

Campbell, B. (2015, February 4). *Senate hearing, Queensland Government Administration related to commonwealth government affairs*. Australia: Parliament of Queensland.

Carvalho, A. (2005). Representing the politics of the greenhouse effect: Discursive strategies in the British media. *Critical Discourse Studies, 2*(1), 1–29.

Carvalho, A. (2007). Ideological cultures and media discourses on scientific knowledge: Re-reading news on climate change. *Public Understanding of Science, 16*(1), 223–243.

Christensen, G. (2015, August 19). Speech at House of Parliament, Hansard.

Cripps, A. (2016, August 30). Speech in Queensland Legislative Assembly, Australia.

COP (2015) Paris Agreement United nations. *United Nations Treaty Collect*, 1–27.

Denniss, R., Adams, P., Campbell, R., & Grudnoff, M. (2016). Never gonna dig you up! COP 21(2015) Paris agreement. Retrieved from http://unfccc.int/files/essential_background/convention/application/pdf/english_paris_agreement.pdf

European Environment Agency. (2017). Climate change, impacts and vulnerability in Europe 2016: An indicator-based report. EEA Report No 1, Luxembourg. Publications Office of the European Union.

Fairclough, N. (1995) Media Discourse. London: Edward Arnold.

Foucault, M. (1981 [1970]). The order of discourse. Inaugural lecture at the Collège de France 2nd December 1970. In Young R. (ed.) Untying the text, A post-structuralist reader. Boston: Routledge and Kegan Paul.

Foucault, M. (1984) "Truth and Power," in Paul Rabinow (ed.), *The Foucault Reader* Pantheon, New York. 51–75.

Fitzgibbon, J. (2004). Sustainable development of Australia's petroleum resources. *The APPEA Journal, 44*(2), 34–38.

Foxwell-Norton, K., & Lester, L. (2017). Saving the Great Barrier Reef from disaster, media then and now. *Media, Culture & Society, 39*(4), 568–581.

Frydenberg, J. (2015, November 30). *Questions without notice—Resource projects in the electorate of Dawson*, Australia: House of Representatives.

Giroux, H., & Bhattacharya, D. (2017). Anti-politics and the scourge of authoritarianism. *Social Identities, 23*(5), 503–517.

Hansard. (2015). Senator Brandis response to Senator Waters. Retrieved from http://parlinfo.aph.gov.au/parlInfo/search/display/display.w3p;query=Id%3A%22chamber%2Fhansards%2Fc3d90ba0-0cf0-40f5-b619-9d3ed6ad10bf%2F0099%22

Harsin, J. (2015). Regimes of posttruth, postpolitics, and attention economies. *Communication, Culture & Critique, 8*(2), 327–333.

Hunt, E. (2017, February 13). Scott Morrison and Ray Hadley laugh about coal prop: 'Great stunt'. The Guardian. https://www.theguardian.com/australia-news/2017/feb/13/scott-morrison-and-ray-hadley-laugh-about-coal-prop-483 great-s.

Jowett, G. S., & O'Donnell, V. (2005). *Propaganda & persuasion*. New York, NY: Sage.

Kirsch, S. (2010). Sustainable mining. *Dialectical anthropology, 34*(1), 87–93.

Landry, M. L. (2015, August 20). Speech by House of Representatives for Capricornia, Queensland at House of parliaments, Hansard.

Lynham, A. (2016, March 15). *Record of proceedings*. Queensland Parliament, Australia.

Macdonald, I. (2016, May 2). Speech at Queensland Parliament, Australia.

Met Office. (2017, January 18). 2016: One of the warmest two years on record. Retrieved fromhttp://www.metoffice.gov.uk/news/releases/2017/2016-record-breaking-year-for-global-temperature

Murphy, K. (2017, February 9). Scott Morrison brings coal to question time: What fresh idiocy is this? *The Guardian*. Retrieved from https://www.theguardian.com/australia-news/2017/feb/09/scott-morrison-brings-coal-to-question-time-what-fresh-idiocy-is-this

Palaszczuk, A. (2016, April 19). Speech by Queensland Premier Anastazia Palaszczuk in the Queensland Assembly, Australia.

Pearse, G. (2009). Quarry vision: Coal, climate change and the end of the resources boom. *Quarterly Essay, 1*(*33*), 1–122.

Potter, W. (2011). *Green is the new red: An insider's account of a social movement under siege*. San Francisco, CA: City Lights Publishers.

Sauvant, A., Fitzgibbon W., & Nurse A. (2016). Be afraid, be very afraid of … the environmental activist? Ideological war, Coercive justice, and Orwellian dystopia. In G. Potter, A. Nurse, & M. Hall (Eds.), *The geography of environmental crime: Palgrave studies in green criminology*. London: Palgrave Macmillan. 59–88.

Sawyer, S. (2010). Human energy. *Dialectical anthropology, 34*(1), 67–75.

Slezak, M. (2016, June 7). The Great Barrier Reef: A catastrophe laid bare. *The Guardian*. Retrieved from https://www.theguardian.com/environment/2016/jun/07/the-great-barrier-reef-a-catastrophe-laid-bare

Taithe, B., & Thornton, T. (Eds.). (1999). *Propaganda: Political rhetoric and identity, 1300–2000*. Oxford, UK: Sutton Publishing.

Taylor, C., and Meinshausen, M. (2014) Joint Report to the Land Court of Queensland on "Climate Change—Emissions." Brisbane, 2014.Retrieved from http://envlaw.com.au/wp-content/uploads/carmichael14.pdf

Van Dijk, T. A. (1993). Principles of critical discourse analysis. *Discourse & society*, *4*(2), 249–283.

Woronov, T. (2017). Waging 'lawfare': Law, environment and depoliticization in neoliberal Australia. *Capitalism, Nature, Socialism*. doi:10.1080/10455752.2017.1385639.

3. *Global Similarities and Persistent Differences: A Survey of Comparative Studies on Climate Change and Communication*

James Painter and Mike S. Schäfer

Introduction: A Comparative Perspective on Climate Change Communication

Climate change is a phenomenon that transcends national boundaries and has been interpreted as a symbol of a 'world risk society' (Beck, 2008, p. 81ff.). Climatic changes are observable on all continents (IPCC, 2014) and have social, economic and political impacts around the globe (e.g. Dryzek et al., 2011).

Climate change is also perceived, interpreted, and contextualized differently in different countries; the global phenomenon is 'domesticated' in different ways (Eide & Kunelius, 2010). The news media play an important role in this process. Due to climate change being a complex, large-scale, invisible process and, thus, an 'unobtrusive' issue for many (Moser, 2010), news media are crucial sources for many peoples' information about climate change. Their coverage on climate change plays an important role in shaping the context within which different publics are made aware of climate change, and are prepared to take action to combat it (e.g. Carvalho, 2010, p. 172).

Due to the importance of news media and the global yet differentiated importance of climate change in different countries, international comparisons of media coverage of climate change are of considerable interest. Yet such studies are not that common. Even though research on climate change communication has expanded considerably in recent years (Moser, 2016; Schäfer & Schlichting, 2014), comparative analyses make up only a fifth of available studies, and their proportion has decreased over time (Schäfer & Schlichting, 2014, 152ff.).

In addition, there are significant gaps in our understanding of the similarities and differences between countries. Scholarship focuses on 'Western', mostly Anglophone countries, and has neglected African, Latin American, and many Asian countries (Schäfer & Schlichting, 2014; for some exceptions see Eide, Kunelius, & Kumpu, 2010; Schmidt, Ivanova, & Schäfer, 2013; Shanahan, 2009). Moreover, due to different analytical foci as well as varying data, analyzed periods of time, and research methods, comparing results across studies is difficult as the 'functional equivalence' (Esser, 2010) of measurements across contexts is often lacking.

The aim of this chapter, therefore, is to identify similarities and differences between countries and regions as well as research gaps, focusing on media representations.

Global Similarities in Climate Change Reporting

Cross-national, comparative research on climate change communication has focused on many facets (for an overview see Schäfer, 2015): the degree of attention media devote to the issue (e.g. Schmidt et al., 2013), the events triggering such coverage (Barkemeyer et al., 2017; Liu, Lindquist, & Vedlitz, 2011; Schäfer, Ivanova, & Schmidt, 2014), the sources journalists quote (Eide et al., 2010; Painter et al., 2016), and the evaluation and framing of the issue including 'sceptical' framing (O'Neill, Williams, Kurz, Wiersma, & Boykoff, 2015; Painter, 2011; Painter & Ashe, 2012). These studies have shown differences and similarities between the analyzed countries.

Firstly, and notwithstanding country differences, scholarship has shown that *climate change has received a considerable amount of media coverage globally* since the mid-2000s. In most countries, the proportion of climate-related news articles has increased since 2000–2005, with peaks occurring between 2007 and 2009 (depending on the country), and around COP21 in 2015 (see Figure 3.1 and Gifford et al., 2017; Schmidt et al., 2013). Also, climate change has become a relevant topic in many countries compared to other issues. A 27-country analysis of media attention for climate change which used similar methodology on countries from all continents in order to ensure equivalent measurements established that climate change was mentioned in between 0.28% (India) and 1.42% (Australia) of all published print media articles between 1997 and 2009 (Schmidt et al., 2013). These figures rose to between 0.54% (Russia) and 3.61% (again Australia) from 2006 to 2009.

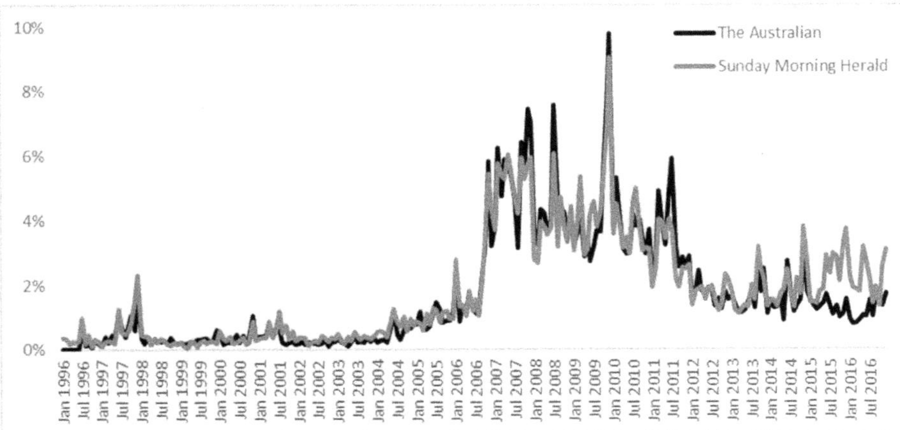
The Australian
Sunday Morning Herald
10%
8%
6%
4%
2%
0%
Jan 1996
Jul 1996
Jan 1997
Jul 1997
Jan 1998
Jul 1998
Jan 1999
Jul 1999
Jan 2000
Jul 2000
Jan 2001
Jul 2001
Jan 2002
Jul 2002
Jan 2003
Jul 2003
Jan 2004
Jul 2004
Jan 2005
Jul 2005
Jan 2006
Jul 2006
Jan 2007
Jul 2007
Jan 2008
Jul 2008
Jan 2009
Jul 2009
Jan 2010
Jul 2010
Jan 2011
Jul 2011
Jan 2012
Jul 2012
Jan 2013
Jul 2013
Jan 2014
Jul 2014
Jan 2015
Jul 2015
Jan 2016
Jul 2016

(a)

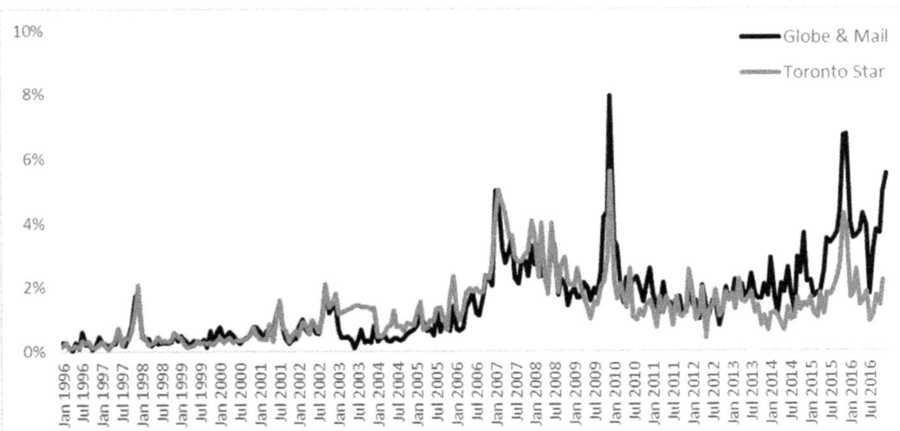
Globe & Mail
Toronto Star
10%
8%
6%
4%
2%
0%
Jan 1996
Jul 1996
Jan 1997
Jul 1997
Jan 1998
Jul 1998
Jan 1999
Jul 1999
Jan 2000
Jul 2000
Jan 2001
Jul 2001
Jan 2002
Jul 2002
Jan 2003
Jul 2003
Jan 2004
Jul 2004
Jan 2005
Jul 2005
Jan 2006
Jul 2006
Jan 2007
Jul 2007
Jan 2008
Jul 2008
Jan 2009
Jul 2009
Jan 2010
Jul 2010
Jan 2011
Jul 2011
Jan 2012
Jul 2012
Jan 2013
Jul 2013
Jan 2014
Jul 2014
Jan 2015
Jul 2015
Jan 2016
Jul 2016

(b)

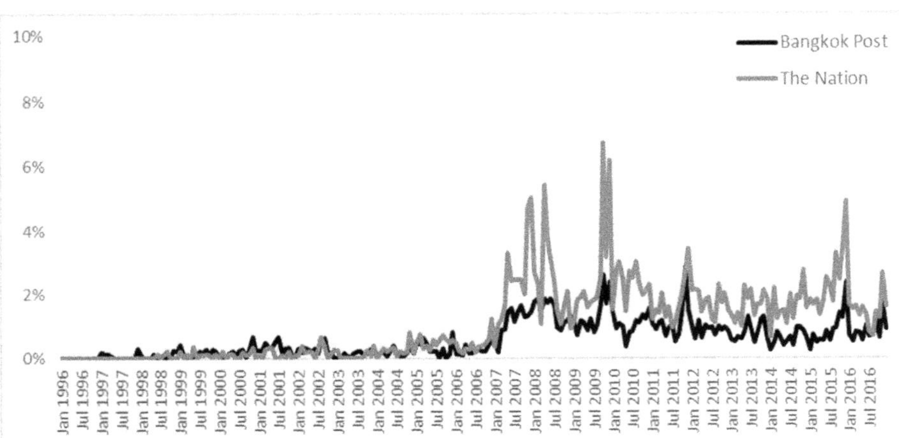
Bangkok Post
The Nation
10%
8%
6%
4%
2%
0%
Jan 1996
Jul 1996
Jan 1997
Jul 1997
Jan 1998
Jul 1998
Jan 1999
Jul 1999
Jan 2000
Jul 2000
Jan 2001
Jul 2001
Jan 2002
Jul 2002
Jan 2003
Jul 2003
Jan 2004
Jul 2004
Jan 2005
Jul 2005
Jan 2006
Jul 2006
Jan 2007
Jul 2007
Jan 2008
Jul 2008
Jan 2009
Jul 2009
Jan 2010
Jul 2010
Jan 2011
Jul 2011
Jan 2012
Jul 2012
Jan 2013
Jul 2013
Jan 2014
Jul 2014
Jan 2015
Jul 2015
Jan 2016
Jul 2016

(c)

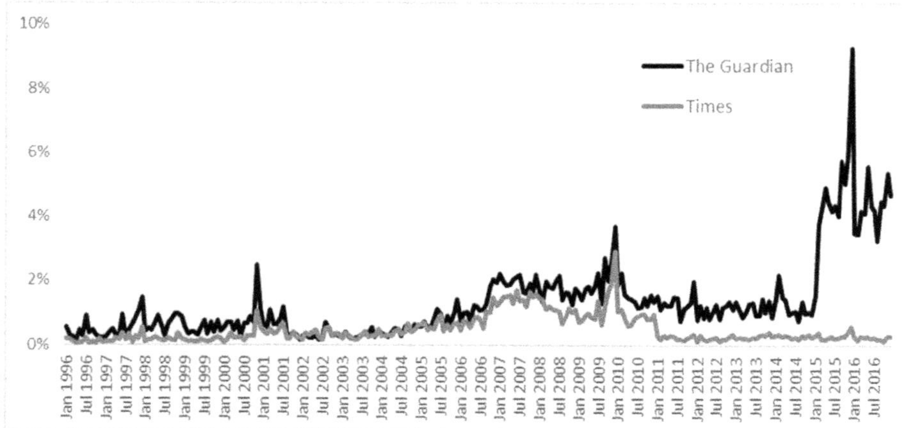
10%
8%
6%
4%
2%
0%
The Guardian
Times

(d)

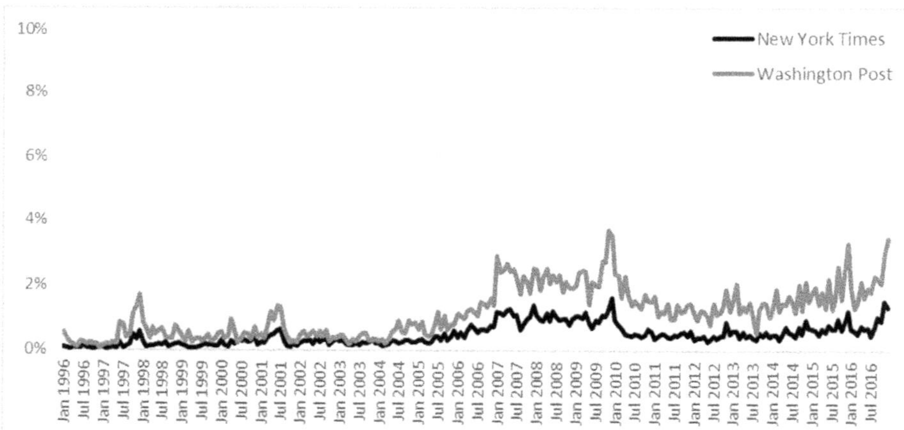
10%
8%
6%
4%
2%
0%
New York Times
Washington Post

(e)

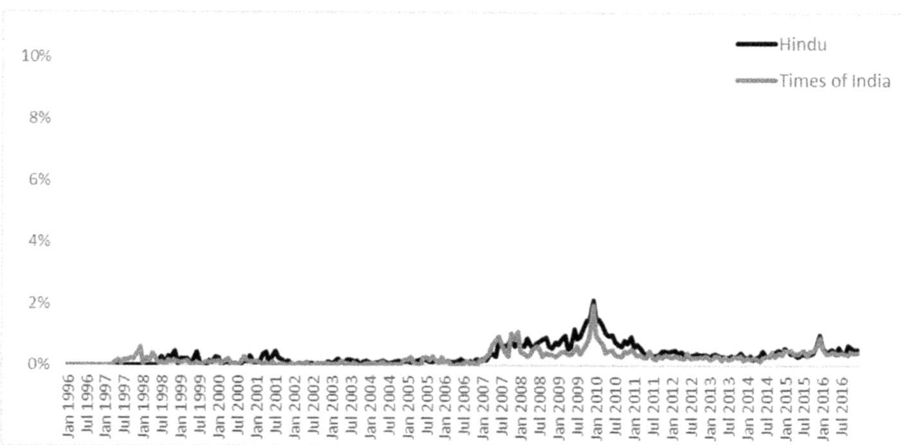
10%
8%
6%
4%
2%
0%
Hindu
Times of India

(f)

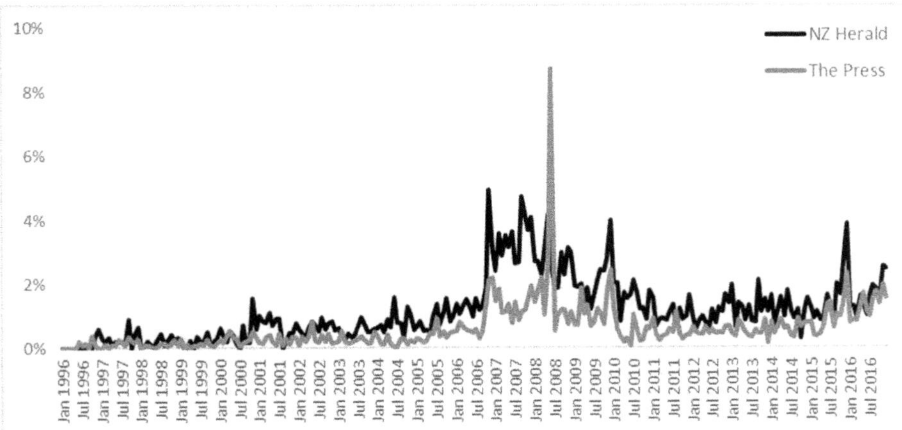

(g)

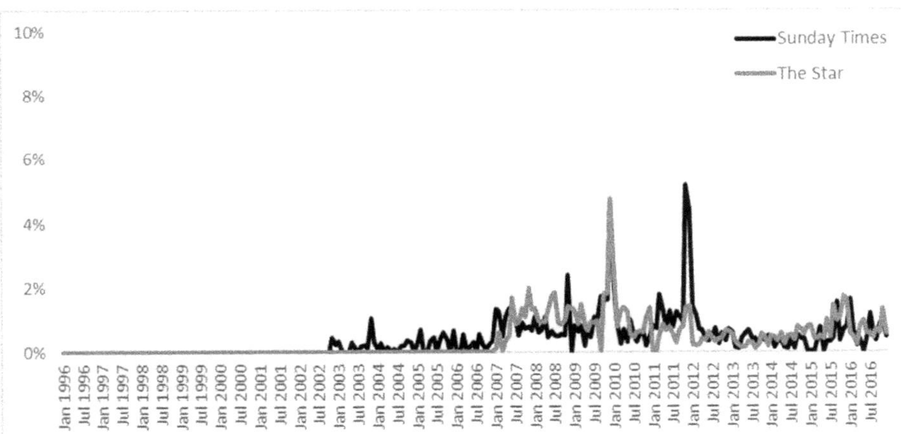

(h)

Figure 3.1: Media attention for climate change from 1996 to 2016 in Australia, Canada, Thailand, the UK, the US, India, New Zealand, and South Africa.

Note: Percentages indicate the amount of articles dealing with climate change (i.e. containing one of the keywords 'climate change', 'global warming' or 'greenhouse effect', which are the most widely used keywords in studies on media attention (Schmidt et al., 2013), at least once) in relation to all articles published in the analyzed newspapers per month.

These percentages outnumber media coverage about other controversial scientific issues: Stem cell research, for instance, received 0.12% of media attention in Germany between 1997 and 2003, or the sequencing of the human genome received 0.06% in Germany (1993–2003), 0.08% in France (1999–2001) and 0.1% in the US (1999–2001, calculations based on Gerhards & Schäfer, 2006, p. 93ff.; Schäfer, 2007, p. 88, 2009).

Secondly, studies have established that *media coverage is strongly event-driven and episodic, and that it focuses on similar events across countries.* Synchronized peaks in media coverage occur during selected Conferences of the Parties (COPs) to the United Nations Framework Convention on Climate Change (UNFCCC) and the publications of the Assessment Reports (ARs) of the IPCC, the Intergovernmental Panel on Climate Change (Eide et al., 2010; Gifford et al., 2017; Schmidt et al., 2013). Among the COPs, COP15 in Copenhagen in 2009, where the formulation of a successor agreement to the Kyoto Protocol was expected but ultimately fell short, figured prominently in media coverage, as did COP21 in Paris 2015 when an agreement was eventually reached (Painter et al., 2016). Climate-related media attention peaked around both events in practically all countries (see Figure 3.1), and time-series analysis also found COPs to be systematic triggers of media coverage on climate change in Australia, India, and Germany (Schäfer et al., 2014).

Thirdly, many studies analyzing media coverage of climate change have focused on the 'framing' of the issue. They have analyzed under which perspective the issue is presented in the media, which of its aspects are highlighted, as well as which claims, demands and responsibilities are deduced from it (for overviews see Nisbet, 2009; O'Neill et al., 2015). Comparative framing analyses have shown that the *spectrum of frames prevalent in different countries displays many similarities.* In their review of previous studies on framing climate change, O'Neill et al. (in the online appendix of O'Neill et al., 2015) show that the frames found in seven previous studies can be integrated into a generic set of ten climate change issue frames ranging from 'settled science' and 'uncertain science' frames over 'disaster' and 'opportunity' to 'health' and 'morality and ethics' frames (see also Painter et al., 2016).

In addition, empirical studies indicate that the relative prevalence of most frames in different Western countries is similar, particularly when coverage is driven by 'events' such as the publication of IPCC reports or COP meetings. For example, a six-country study analyzing newspaper coverage of the IPCC's 4th AR, the 2012 IPCC report on extreme weather events, and Arctic sea ice melt in Australia, France, India, Norway, UK, and the US showed that the distribution of frames was similar across countries, with the 'disaster/implicit risk' and 'uncertainty' frames clearly dominating (Painter, 2013).

Print media coverage of the IPCC's 5th AR was also framed similarly in five developed (Canada, France, Japan, Norway, and the UK) and five non-developed countries (Bangladesh, Brazil, Chile, Indonesia, and South Africa) (Painter, 2017). Six of seven themes—disaster, uncertainty, skepticism, risk, opportunity of action and opportunity of inaction—occurred similarly often in these countries' print media. The lack of cross-country differences was also found in television coverage of the AR5, where disaster and uncertainty were again much more prominent than risk and opportunity in Australia, Brazil, China, Germany, India, and the UK (Painter, 2014). Visual framing in print coverage seems to be similar, as well, across the largest English- (O'Neill, 2013; O'Neill, Boykoff, Niemeyer, & Day, 2013) and German-speaking countries (Metag, Schäfer, Füchslin, Barsuhn, & Kleinen-von Königslöw, 2016, p. 199 ff.), where climate change is visually presented using disaster framing and 'talking heads'.

An analysis of print media content focusing on COPs 16 to 19 also concluded that the textual and visual framing of climate change and climate politics in Brazil, Germany, India, South Africa, and the United States is strongly convergent between the countries even though these frames are subsequently embedded into different national narratives (Lück, Wessler, Wozniak, & Lycarião, 2016).

Fourthly, *there seems to be a 'societal turn' in media coverage on climate change*, that is, a move away from debating the underlying science of climate change towards a debate of appropriate political measures and societal implications. Several indicators support this assumption, even though the evidence is more scarce here compared to the previous points. But content analysis analysing media coverage over long time-spans in Germany have shown that climate change has moved from the science sections of the media, where it was mostly placed in the 1990s and early 2000s, into the politics, economy and even culture sections (Ivanova, 2012) (see Figure 3.2). Computer-assisted analyses of US media also provide evidence that words indicating scientific aspects of climate change have been used less often over time, whereas words indicating political or economic aspects have risen in importance (Kirilenko & Stepchenkova, 2012; Kirilenko, Stepchenkova, Romsdahl, & Mattis, 2012). A shift away from scepticism about the science of climate change towards a scepticism surrounding the politics of climate change has been shown in the US as well (Schmid-Petri, Adam, Schmucki, & Häussler, 2015).

An analysis of print media coverage on climate change using corpus-linguistic methods for 11 countries including Australia, Canada, Germany, India, Namibia, New Zealand, Singapore, South Africa, Thailand, the UK, and the US (Ivanova, 2017, p. 263ff.) reported similar results. It seems that while 'the scientific basis of climate change is still being reported upon at times (and, especially in Anglo-Saxon countries, sometimes debated; Painter & Ashe, 2012), aspects like the political efforts towards a global treaty on emissions reductions, the costs of moving towards low carbon societies for national economies, or the humanitarian and security implications connected to potentially intensifying resource conflicts and climate-induced migration, are increasingly taking center stage' (Schäfer, 2015, p. 856ff.).

Persistent Country Differences

Despite these similarities in international media coverage, there are also persistent differences between countries in the volume of coverage, the amount of attention given to sceptics, the emphasis on different themes (away from set events or climate reports), and the degree of 'domestication'. These differences often lie between 'Anglosphere' countries, continental European countries and countries in the 'Global South'.

Country disparities in the volume of media coverage over time, which is often described as 'issue attention', have been monitored in different studies. A study of AR5 print coverage in 22 countries in 2013–2014 showed differences between (i) a group of industrialised, usually Anglosphere, countries (Australia, Canada, Finland, Japan, USA, and the UK, who published around 90 or more articles), (ii) a group of continental European countries (France, Germany, Norway, and Sweden who published 40 to 80 articles), and (iii) a group from the Global South (Bangladesh, Chile, China, Egypt, Ethiopia, Indonesia, South Africa, and Uganda, who published between 5 and 40 articles) (Kunelius, Eide, Tegelberg, & Yagodin, 2017, 21).

Schmidt et al. (2013) have tried to ensure a greater 'functional equivalence' in measuring issue attention between countries. They used comparable data from 27 countries from 1996 to 2010, and found issue attention (measured by the mean percentages of the total articles in a newspaper dedicated to climate change) high in Australia, Canada, Indonesia, and the UK, relatively high in France, Ireland, New Zealand, and the US, and relatively low or low in continental European countries (Germany, the Netherlands, and Spain) and various countries in the Global South (including Brunei, India, and Mexico; see also Figure 3.1 for a continuation of similar data for eight countries).

Similar results emerged from another wide-ranging study, this time of 2.6 million articles in 32,000 newspaper issues across 113 newspapers published in 41 countries (Barkemeyer et al., 2017) in one year—2008. Again, Australia, Canada, and the UK had the highest volume of coverage at an average of more than two articles per newspaper issue. This was followed by a second group of countries at between one and two articles (China, Germany, India, Ireland, Mexico, and New Zealand). The US was relatively high at 0.89 articles per issue. The lowest were Venezuela, Guatemala, Ecuador, Costa Rica, and Russia at less than one article for every ten issues.

Less clear than these volume differences are their *drivers—but also here, country differences are observable.* Rooted in agenda-setting theory, multiple streams theory and punctuated equilibrium theory, Schäfer et al. (2014) showed that weather and climate characteristics are not very strong drivers of media coverage in Australia, Germany, and India, and that international socio-political events like the COPs or domestic political debates and contestation trigger media coverage on climate change more strongly.

Generally, it seems clear that one of the main drivers of differences in issue attention is when aspects of climate change—the science explaining it, its impacts or the policy responses to it—are heavily contested in the political sphere. For example, in Australia, between June 2010 and September 2015, disputes over climate change policy played a significant causative role in the changing or removal of five prime ministers over this period (cf. Bacon & Nash, 2012). So high political contestation often translates into high media coverage of climate change, particularly as the media often take their cues from politicians and quote politicians as one of their main sources, even on the science (Painter, 2010). Indeed, Australia is one of a group of Anglosphere countries, along with Canada, the US, and the UK, where extensive media analysis has shown that various forms of scepticism, often held by right-wing politicians, is common, particularly in the opinion sections of the right-leaning legacy media (Elsasser & Dunlap, 2013; Painter & Gavin, 2016).

A third area of difference between countries can be found in the presence or absence of different themes or frames. As we have seen, the *spectrum* of themes is often similar across countries, but differences emerge from longitudinal studies focusing on the *frequency* of specific frames. We know that scepticism is more prominent in Anglosphere countries (Grundmann & Scott, 2014; Painter, 2011; Painter & Ashe, 2012) and less prominent in continental Europe and developing countries like China and India (Billett, 2010; Dispensa & Brulle, 2003; Engels, Hüther, Schäfer, & Held, 2013). Other studies have found, for example, that the presentation of climate change as a security issue differs markedly between Western countries, who strongly focus on national and energy security, and developing countries who place greater emphasis on food and water security (Schäfer, Scheffran, & Penniket, 2015). The Indian media exhibit some marked differences to those in developed countries. Billett (2010) showed from a study of the English language print media between 2002 and 2007 that the dominant frames were strongly shaped by a post-colonial and developmental approach rooted in India's vulnerability to climate risks and a desire for a just distribution of responsibilities between developed and developing countries.

Finally, *research has shown important differences in the degree of 'domestication' of climate change*, that is, in the degree to which the media in different countries 'domesticate global climate politics by opening up national spaces of interpretation' (Kunelius & Yagodin, 2016, p. 84). These authors' study of AR5 print media coverage in 22 countries showed a wide range of country differences, when it mapped how scientists, politicians, NGO representatives, businessmen and women, and other sectors appeared as sources in the media. China, Finland, France, Germany, Japan, Poland, Russia, South Africa, and Sweden were classified as exhibiting 'high domestication', Australia, Brazil, Canada, Norway, the UK, and the US as 'medium domestication', and Bangladesh, Chile, Egypt, Ethiopia, Indonesia, Israel, and Uganda as 'low domestication'. Two studies focusing on the specific differences between the media in Germany and the US have found that foreign voices are stronger in Germany, whereas domestic voices dominate in the US (Ivanova, 2017; Konieczny, 2013).

Country differences seem easier to map than to explain. No definitive study has come up with a robust, commonly-applicable framework for analyzing the main drivers of these differences. Researchers have chosen different variables, specific to the media and journalism in discrete countries, or wider societal factors, as the main explanatory forces behind the country differences. For example, Schmidt et al. (2013) chose two 'societal' variables to assess a correlation with media attention, namely a country's vulnerability to climate risks and its carbon dependency. They found firstly, that anticipated climate risks did correlate with higher media attention in countries not bound by the Kyoto Protocol (i.e., non-Annex B), mostly found in the Global South; and secondly, that the extent of media attention in Annex B countries corresponded with the degree of carbon dependency. The paradigmatic case which supported the second hypothesis was Australia, which had high media attention and high carbon dependency. Most European countries in their study, by contrast, showed a lower carbon intensity of GDP and lower levels of media attention. However, there were some obvious, but significant, outliers. The UK is one of them, exhibiting high media attention but low carbon dependency. Another was Russia, which had low media attention but a very carbon-reliant economy.

Barkemeyer et al. (2017) also select climate vulnerability as a country variable, but add a political dimension (the quality of a country's regulatory regime), economic dimensions (level of economic development, carbon intensity, integration into the world economy), a labour dimension (unemployment levels), and a cultural dimension (distribution of religious denominations). They conclude that there is evidence for a correlation between levels of media coverage and a country's direct exposure to climate change, and the measures taken to combat climate change. But crucially, they argue, a number of contextual factors play a major role, particularly regulatory quality (correlated with high media coverage) and unemployment levels (correlated with low media coverage as concern about high unemployment crowds out coverage of other issues).

These authors seem right to stress the societal context in which climate change reporting takes place. But such studies downplay the internal factors such as 'first order' or 'second order' journalistic norms like balance, dramatization or novelty (Boykoff, 2007; Boykoff & Boykoff, 2004, 2007), editorial culture or norms (Nisbet, Brossard, & Kroepsch, 2003), the influence of proprietors (McKnight, 2012), the role of a media organisation's ideology (Carvalho & Burgess, 2005; Dirikx & Gelders, 2010; Painter & Gavin, 2016), and the resources available to journalists in different countries (Shanahan, 2009). They also leave out other contextual factors which may help to explain why climate change is subject to more political contestation in some countries compared to others (and often as a result, more coverage) such as the presence (or not) of organized scepticism through lobbying groups or think tanks (McCright & Dunlap, 2000; Oreskes & Conway, 2010; Painter, 2011), the influence of fossil fuel companies (Dispensa & Brulle, 2003), the presence of sceptical politicians, readers, or scientists (Painter, 2016), and country-specific cultural or political values (Engels et al., 2013; Kahan, Jenkins-Smith, & Braman, 2011).

The Role of New Online Players in Different Countries

In an increasingly fragmented media landscape in many countries, new 'digital-born' players, such as Huffington Post (HuffPo), Quartz, Vice, Vox, and BuzzFeed, form part of a new digital presence alongside blogs and microblogs, niche sites, social media platforms and legacy media sites making the transition from print versions to online or adding online presences to their existing platforms. Some of the digital-born players are making inroads into the dominance of legacy media, both for news in general and for news about the environment (Fletcher, 2016; Newman, Fletcher, Levy, & Nielsen, 2016). Many of them place a strong emphasis on coverage of the environment, in part because they believe that this area is not covered well enough by legacy media, and is of particular interest to their younger target audience (Painter et al., 2016).

Mostly US-based, some of them are now investing heavily in various languages and discrete versions aimed at consumers in specific countries or regions. They vary greatly in terms of their business models, distribution strategies, and editorial priorities (Nicholls, Shabbir, & Nielsen, 2016). Even though both Huffington Post and BuzzFeed rely on complicated algorithms to push out their content on a variety of social media, HuffPo was founded as a left-leaning news outlet, whereas BuzzFeed is essentially a tech company with a media layer on top (Painter et al., 2016, p. 13). Country differences in content produced by these new players are driven both by these variations and by a diversity of local factors such as the strength of existing legacy media, the relationship between parent company and local partner, the available resources, and the capacity for original journalism.

Some indication of the way digital-born players are different both to legacy players and to each other can be found in a detailed study of the media coverage of the Paris COP summit at the end of 2015 (Painter et al., 2016). For example, sharp differences emerged between legacy and digital-born players in the volume of coverage. The application of the same methodology to select articles from two legacy and three digital-born players in six countries gave a sample size of nearly 2,000 articles across 22 days of analysis. Of these over two-thirds appeared in legacy media, mostly in the left-leaning media. However, there were also sharp differences in the amount of coverage by the new players, and between countries (see Figure 3.2). Huffington Post in English and French stood out for the large number of articles. Indeed, in the UK, US, and France, Huffington Post's coverage was larger than that of the right-leaning media organisation selected. In the case of the US, it had the most volume of any outlet—more than double that of the New York Times. By contrast, in Spanish and German the volume was significantly less than legacy media.

Various explanatory factors come into play. In the US, Huffington Post has identified a gap in environmental coverage, particularly for its target

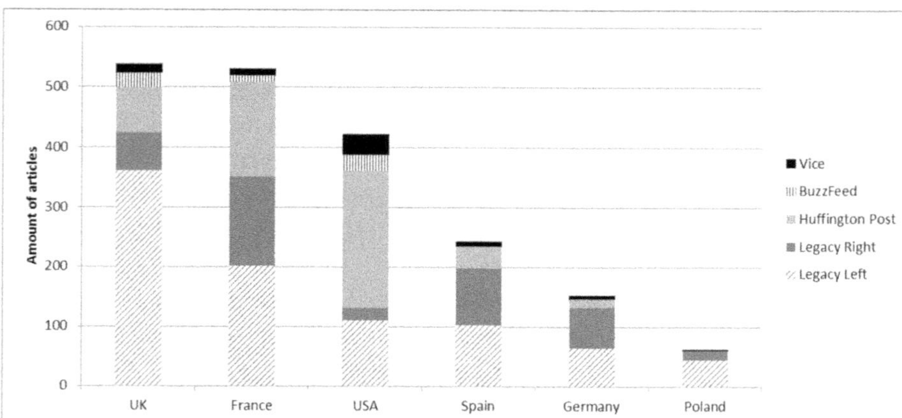

Figure 3.2: Volume of coverage of COP21, by country and media type/organisation, 25/11/2016 to 16/12/2016 (many thanks to Dr. Silje Kristiansen).

Source: Adapted from Painter et al. (2016).

audience, in part caused by the crisis affecting the business models of legacy media there (Benson, 2016). However, the high volume is also a product of its business model which is driven by the imperative of publishing 'volume' so as to maximise the various types of advertising possibilities (native, display, video, programmatic) around the content. In France, another factor is that media in countries hosting COPs traditionally publish more than other countries due to the easier deployment of resources, the expectation of its audiences, and the ready availability of national politicians and other sources.

BuzzFeed and Vice publish much less content, in part because they specifically aim to be different from legacy media by selecting material or formats which makes them stand out and/or is shareable on social media. There are important country differences too: in Poland, Spain, and Germany, there was very little coverage of the summit, even though there are versions in those languages.

Probably the most important explanatory factor here are resources, as these countries have significantly fewer journalists working for these new sites compared to their parent sites in English (Painter et al., 2016). This may change if the new players achieve greater consolidation in different countries.

Huffington Post again is a different case, as they have reasonably sized newsrooms in France (around 30 full-time staffers in 2016), Germany (21) and Spain (15)—but significantly less than the US (260) and the UK (40). In the first three countries, one of the many drivers of editorial output is Huffington Post's relationships with its partner organisations, including the degree of collaboration or the desire for a distinct identity, and how much material is simply translated from the versions in English.

As regards country differences in the themes that the new players focused on, all three of them in most countries tended to cover the climate negotiations significantly less than the legacy media (see Figure 3.3). Another difference is that all of them virtually ignored scepticism except to make fun of it. This contrasted with the scepticism found in the right-leaning legacy media in the *Telegraph* in the UK, *USA Today*, and *Le Figaro* in France. In the US, UK, and Germany, the Huffington Post placed considerable emphasis on reporting the opportunities opened up by climate change (such as green jobs in renewable energy), which is very much in line with the 'solutions' journalism promoted by its former owner, Arianna Huffington, and the approach offered by its editors who say they are 'passionate about content which advertisers find attractive, including solution-based reporting and issues such as parenting or sustainability' (Nicholls et al., 2016, p. 17).

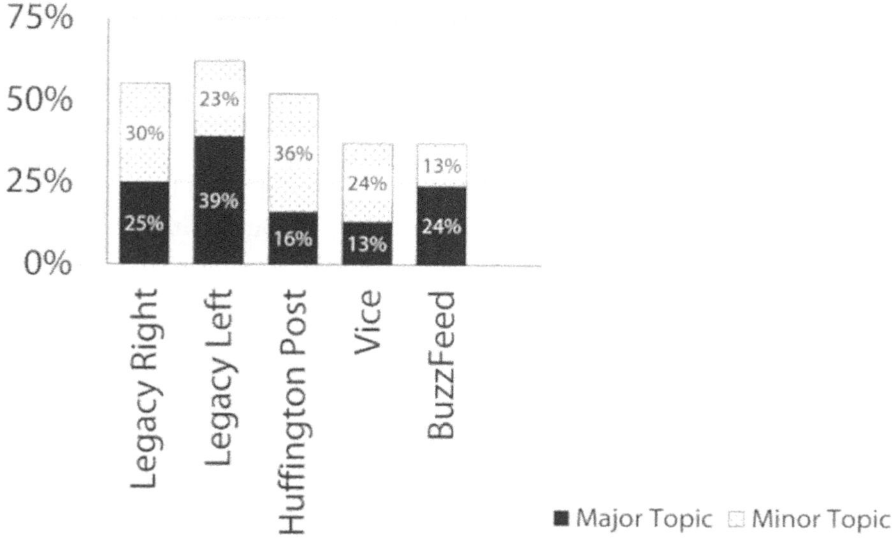

Figure 3.3: Percentage of articles by media type/organisation covering the negotiations, as major and minor topic.

Source: Painter et al. (2016).

Conclusions

Climate change is global in character and in its impacts, and the attempts to alleviate the negative effects of climate change have often been located at an international or global level—epitomized by the annual climate change summits. Therefore, focusing scholarly attention on international comparisons of media coverage of climate change seems important, yet only few such studies have been published and significant gaps in our understanding of the similarities and differences between countries exist. We have summarized the main findings of comparative studies of climate change communication here, exemplifying them with our own research.

We were able to show that, on the one hand, a number of important cross-national parallels exist in climate change coverage: climate change has become a global media issue, where coverage is often event-driven and focuses particularly on global socio-political events like the COPs. The spectrum of interpretative frames present in different nations is relatively similar. And in many countries, a 'societal turn' seems to have occurred in which coverage turned away from the scientific aspects of climate change towards the political and social ones.

On the other hand, scholarship also points out persistent differences between countries. These regard the amount of space given to sceptical voices, the overall volume of climate change coverage in different countries, and the relative frequency of dominant frames used to interpret the topic other than in the coverage of set events or reports.

In addition, new digital-only players, such as Huffington Post, Vice, and BuzzFeed, are making inroads into the dominance of legacy media with a strong emphasis on coverage of the environment and a growing presence in a variety of languages and countries. But their relative strength and importance also varies between countries quite significantly. Overall, our synopsis demonstrates the fruitfulness of a cross-nationally comparative perspective as well as the many limitations and gaps in current scholarship.

More comparative studies are needed, particularly ones comparing climate change coverage with (or between) countries in the Global South, with emerging economies such as China, India, or Brazil (Lück et al., 2016; Painter, 2011), or focusing on the most vulnerable countries to climate change which are exclusively situated in the global south (Schäfer & Schlichting, 2014). For example, given the growing attention drawn to health risks associated with a changing climate such as a possible rise in human infectious diseases (Wu, Zu, Zhou, Chen, & Xu, 2016), changing consumption patterns of fruit and vegetables (Springmann et al., 2016) or greater levels of air pollution (Hassan, Hashim, & Hashim, 2016), we do not know if such risks are given more prominence in the media in developing countries where the impacts are more likely to be felt. One study of the US media showed that the number of print articles about public health risks rose markedly between 2011 and 2012 compared to 2007–2008 (Weathers & Kendall, 2016), but scholarship is lacking about the prominence of this theme in other countries. A similar point could be made about other aspects of the climate change 'story' such as extreme weather events, migration and displacement, or the causes of international disputes, and their possible links to climate change.

Furthermore, as print newspapers are in crisis in many countries, we need more analyses of TV—which is the main source of climate change-related information for many—and on online and social media, which are rising in importance. Such a shift would have to be accompanied by a focus on different modalities of climate change coverage, for example, a focus on visuals and videos which are found in abundance in digital players like Vox and Quartz (Painter et al., 2016), and also on alt-right sites like Breitbart.

(Re)assessing temporal trends that have been described in recent years between countries would also be worthwhile, especially in light of the rise of nationalist and/or populist movements in many countries which often advance scepticism towards climate change or display hostility towards institutionalized science or scientific 'experts'.

Finally, as we have seen above, the relative weighting of factors internal to journalism and of contextual societal factors influencing media debates on climate change is difficult to pin down to explain country differences or similarities. This is due to the enormous variety of factors influencing journalists' work in different countries, and the huge range of contextual factors evidenced by national economic, social, and cultural indicators (Kunelius et al., 2017, pp. 85–88). A good starting point for identifying the influences on the journalistic process can still be found in the work by Shoemaker and Reese (1996), which describes concentric circles ranging from 'the micro-scale of individual attributes, to meso-influences of professional norms, work routines, and organizational cultures, to the macro-influences of economic imperatives, ownership structures, and cultural/ideological values' (Gibson, 2017). So for example, in our understanding of why climate scepticism in common in the media of Anglosphere countries but not of others, such an approach could, as a replicable methodology, give insights into the relative strength of the roles played by editors, media owners, and individual journalistic values on the one hand, and the role of lobby groups, fossil fuel companies, skeptical politicians, and readers' values on the other.

References

Bacon, W., & Nash, C. (2012). Playing the media game: The relative (in)visibility of coal industry interests in media reporting of coal as a climate change issue in Australia. *Journalism Studies, 13*(2), 243–258. doi:10.1080/1461670x.2011.646401

Barkemeyer, R., Figge, F., Hoepner, A., Holt, D., Kraak, J. M., & Yu, P.-S. (2017). Media coverage of climate change: An international comparison. *Environment and Planning C: Politics and Space,* doi:10.1177/0263774X16680818

Beck, U. (2008). *World at risk.* Cambridge: Polity.

Benson, R. (2016). Are foundations the solution to the American journalistic crisis?. *NYU Media Ownership Project Working Paper, 1.*

Billett, S. (2010). Dividing climate change: Global warming in the Indian mass media. *Climatic Change*, *99*(1–2), 1–16.

Boykoff, M. T. (2007). Flogging a dead norm? Newspaper coverage of anthropogenic climate change in the United States and United Kingdom from 2003 to 2006. *Area*, *39*(2), 470–481.

Boykoff, M. T., & Boykoff, J. M. (2004). Balance as bias: Global warming and the US prestige press. *Global Environmental Change, 14*(4), 125–136.

Boykoff, M. T., & Boykoff, J. M. (2007). Climate change and journalistic norms: A case-study of U.S. mass-media coverage. *Geoforum*, *38*(6), 1190–1204.

Carvalho, A. (2010). Media(ted) discourses and climate change: A focus on political subjectivity and (dis)engagement. *WIREs Climate Change, 1*(2), 172–179. doi:10.1002/wcc.13

Carvalho, A., & Burgess, J. (2005). Cultural circuits of climate change in U.K. broadsheet newspapers, 1985–2003. *Risk Analysis, 25*(6), 1457–1469.

Dirikx, A., & Gelders, D. (2010). Ideologies overruled? An explorative study of the link between ideology and climate change reporting in Dutch and French newspapers. *Environmental Communication, 4*(2), 190–205.

Dispensa, J., & Brulle, R. (2003). Media's social construction of environmental issues: Focus on global warming—A comparative study. *The International Journal of Sociology and Social Policy, 23*(10), 74–105.

Dryzek, J. S., Norgaard, R. B., & Schlosberg, D. (Eds.). (2011). *The Oxford handbook of climate change and society.* Oxford University Press.

Eide, E., & Kunelius, R. (2010). Domesticating global moments. A transnational study on the coverage of the Bali and Copenhagen Climate Summits. In E. Eide, R. Kunelius, & V. Kumpu (Eds.), *Global climate—Local journalisms* (pp. 11–50). Bochum: projektverlag.

Eide, E., Kunelius, R., & Kumpu, V. (Eds.). (2010). *Global Climate—Local journalisms.* Bochum: projektverlag.

Elsasser, S. W., & Dunlap, R. E. (2013). Leading voices in the Denier choir: Conservative columnists' dismissal of global warming and denigration of climate science. *American Behavioral Scientist, 57*(6), 754–776.

Engels, A., Hüther, O., Schäfer, M., & Held, H. (2013). Public climate-change scepticism, energy preferences and political participation. *Global Environmental Change, 23*(5), 1018–1027.

Esser, F. (2010). *Komparative Kommunikationswissenschaft. Working Paper No. 41.* Zürich: National Centre of Competence in Research (NCCR).

Fletcher, R. (2016). The public and news about the environment. In: Painter et al. (Eds.), *Something old, something new: Digital media and the coverage of climate change* (pp. 24–36). Oxford: Reuters Institute for the Study of Journalism.

Gerhards, J., & Schäfer, M. S. (2006). *Die Herstellung einer öffentlichen Hegemonie. Humangenomforschung in der deutschen und der US-amerikanischen Presse.* Wiesbaden: Verlag für Sozialwissenschaften.

Gibson, T. A. (2017). Economic, technological, and organizational factors influencing news coverage of climate change. *Oxford Research Encyclopedias.* doi:10.1093/acrefore/9780190228620.013.355

Gifford, L., Luedecke, G., McAllister, L., Nacu-Schmidt, A., Andrews, K., Boykoff, M., & Daly, M. (2017). *World newspaper coverage of climate change or global warming, 2004–2017.* Retrieved from http://sciencepolicy.colorado.edu/icecaps/research/media_coverage/world/index.html

Grundmann, R., & Scott, M. (2014). Disputed climate science in the media: Do countries matter? *Public Understanding of Science, 23*(2), 220–235.

Hassan, N. A., Hashim, Z., & Hashim, J. H. (2016). Impact of climate change on air quality and public health in urban areas. *Asia Pacific Journal of Public Health, 28*(2S), 38S–48S.

IPCC. (2014). *Climate change 2014. Synthesis report.* Geneva: Intergovernmental Panel on Climate Change.

Ivanova, A. (2012, May). Globale Medienereignisse in der 'globalen MediaPolis'? Struktur und Bestimmungsfaktoren der Ähnlichkeit transnationaler Themenaufmerksamkeit. German Communication Association. 57. Jahrestagung der DGPuK, Berlin.

Ivanova, A. (2017). Transnationalisierung von Öffentlichkeiten. Eine länderübergreifende Längsschnittanalyse der Berichterstattung zum Klimawandel in führenden Tageszeitungen. Wiesbaden: Verlag für Sozialwissenschaften.

Kahan, D. M., Jenkins-Smith, H., & Braman, D. (2011). Cultural cognition of scientific consensus. *Journal of Risk Research, 14*(2), 147–174.

Kirilenko, A., & Stepchenkova, S. (2012). Climate change discourse in mass media: Application of computer-assisted content analysis. *Journal of Environmental Studies and Sciences, 2*(2), 178–191.

Kirilenko, A., Stepchenkova, S., Romsdahl, R., & Mattis, K. (2012). Computer-assisted analysis of public discourse: A case study of the precautionary principle in the US and UK press. *Quality & Quantity, 46*, 501–522.

Konieczny, E. (2013). *Gipfel, Krisen, Konferenzen: Die Entstehung Diskursiver Macht in transnationalen Kommunikationsereignissen*: Springer-Verlag.

Kunelius, R., Eide, E., Tegelberg, M., & Yagodin, D. (Eds.). (2017). *Media and global climate knowledge: Journalism and the IPCC.* New York, NY: Palgrave Macmillan.

Liu, X., Lindquist, E., & Vedlitz, A. (2011). Explaining media and congressional attention to global climate change, 1969–2005: An empirical test of agenda-setting theory. *Political Research Quarterly, 64*(2), 405–419. doi:10.1177/1065912909346744

Lück, J., Wessler, H., Wozniak, A., & Lycarião, D. (2016). Counterbalancing global media frames with nationally colored narratives: A comparative study of news narratives and news framing in the climate change coverage of five countries. *Journalism*, doi:10.1177/1464884916680372

McCright, A. M., & Dunlap, R. E. (2000). Challenging global warming as a social problem: An analysis of the conservative movement's counter-claims. *Social Problems, 47*(4), 499–522.

McKnight, D. (2012). *Rupert Murdoch: An investigation of political power.* Sydney: Allen & Unwin.

Metag, J., Schäfer, M. S., Füchslin, T., Barsuhn, T., & Kleinen-von Königslöw, K. (2016). Perceptions of climate change imagery evoked salience and self-efficacy in Germany, Switzerland, and Austria. *Science Communication, 38*(2), 197–227.

Moser, S. C. (2010). Communicating climate change: History, challenges, process and future directions. *WIREs Climate Change, 1*, 31–53.

Moser, S. C. (2016). Reflections on climate change communication research and practice in the second decade of the 21st century: What more is there to say? *Wiley Interdisciplinary Reviews: Climate Change, 7*(3), 345–369. doi:10.1002/wcc.403

Newman, N., Fletcher, R., Levy, D., & Nielsen, R. K. (2016). *Digital news report.* Oxford: Reuters Institute for the Study of Journalism.

Nicholls, T., Shabbir, N., & Nielsen, R. K. (2016). *Digital-born news media in Europe.* Oxford: Reuters Institute for the Study of Journalism.

Nisbet, M. C. (2009). Communicating climate change: Why frames matter for public engagement. *Environment, 51*(2), 514–518.

Nisbet, M. C., Brossard, D., & Kroepsch, A. (2003). Framing science the stem cell controversy in an age of press/politics. *The International Journal of Press/Politics, 8*(2), 36–70.

O'Neill, S., Williams, H. T. P., Kurz, T., Wiersma, B., & Boykoff, M. (2015). Dominant frames in legacy and social media coverage of the IPCC Fifth Assessment Report. *Nature Climate Change, 5*(4), 380–385.

O'Neill, S. J. (2013). Image matters: Climate change imagery in US, UK and Australian newspapers. *Geoforum, 49*, 10–19.

O'Neill, S. J., Boykoff, M. T., Niemeyer, S., & Day, S. (2013). On the use of imagery for climate change engagement. *Global Environmental Change, 23*(2), 413–421. doi:10.1016/j.gloenvcha.2012.11.006

Oreskes, N., & Conway, E. (2010). *Merchants of doubt: How a handful of scientists obscured the truth on issues from tobacco smoke to global warming.* New York, NY: Bloomsbury Press.

Painter, J. (2010). *Summoned by science: Reporting climate change at Copenhagen and beyond.* Oxford: Reuters Institute for the Study of Journalism.

Painter, J. (2011). *Poles apart: The international reporting on climate skepticism.* Oxford: Reuters Institute for the Study of Journalism.

Painter, J. (2013). *Climate change in the media: Reporting risk and uncertainty.* Oxford: IB Tauris.

Painter, J. (2014). *Disaster averted? Television coverage of the 2013/14 IPCC's climate change reports.* Oxford: Reuters Institute for the Study of Journalism.

Painter, J. (2016). Journalistic depictions of uncertainty about climate change across countries. *Oxford Research Encyclopedias,* doi:10.1093/acrefore/9780190228620.013.346

Painter, J. (2017). Disaster, risk or opportunity? A ten-country comparison of themes in coverage of the IPCC AR5. In: R. Kunelius, E. Eide, M. Tegelberg, & D. Yagodin (Eds.), *Media and global climate knowledge* (pp. 109–128). New York, NY: Palgrave Macmillan.

Painter, J., & Ashe, T. (2012). Cross-national comparison of the presence of climate skepticism in the print media in six countries, 2007–10. *Environmental Research Letters, 7*(4), 1–8.

Painter, J., & Gavin, N. (2016). Climate skepticism in British newspapers, 2007–2011. *Environmental Communication, 10*(4), 432–452.

Painter, J., et al. (2016). *Something old, something new. Digital media and the coverage of climate change.* Oxford: Reuters Institute for the Study of Journalism.

Schäfer, M. S. (2007). *Wissenschaft in den Medien. Die Medialisierung naturwissenschaftlicher Themen.* Wiesbaden: Verlag für Sozialwissenschaften.

Schäfer, M. S. (2009). From public understanding to public engagement: An empirical assessment of changes in science coverage. *Science Communication, 30*(4), 475–505.

Schäfer, M. S. (2015). Climate change in the media. In J. D. Wright (Ed.), *International encyclopedia of the social & behavioral sciences* (2nd ed., Vol. 3, pp. 853–859). Oxford: Elsevier.

Schäfer, M. S., Ivanova, A., & Schmidt, A. (2014). What drives media attention for climate change? Explaining issue attention in Australian, German and Indian Print Media from 1996 to 2010. *International Communication Gazette, 76*(2), 152–176.

Schäfer, M. S., Scheffran, J., & Penniket, L. (2015). Securitization of media reporting on climate change? A cross-national analysis in nine countries. *Security Dialogue, 47*(Online First), 1–21.

Schäfer, M. S., & Schlichting, I. (2014). Media representations of climate change: A meta-analysis of the research field. *Environmental Communication: A Journal of Nature and Culture, 8*(2), 142–160. doi:10.1177/0963662510387559

Schmid-Petri, H., Adam, S., Schmucki, I., & Häussler, T. (2015). A changing climate of skepticism: The factors shaping climate change coverage in the US press. *Public Understanding of Science,* doi:10.1177/0963662515612276.

Schmidt, A., Ivanova, A., & Schäfer, M. S. (2013). Media attention for climate change around the World: A comparative analysis of newspaper coverage in 27 countries. *Global Environmental Change, 23(5), 1233–1248.*

Shanahan, M. (2009). Time to adapt?: Media coverage of climate change in nonindustrialised countries. In T. Boyce & J. Lewis (Eds.), *Climate change and the media* (pp. 145–157). New York, NY: Peter Lang.

Shoemaker, P. J. & Reese, S. D. (1996). *Mediating the message: Theories of influences on mass media content.* New York, NY: Longman.

Springmann, M., Mason-D'Croz, D., Robinson, S., Garnett, T., Godfray, H. C. J., Gollin, D., … & Scarborough, P. (2016). Global and regional health effects of future food production under climate change: A modelling study. *The Lancet, 387* (10031), 1937–1946.

Weathers, M. R., & Kendall, B. E. (2016). Developments in the framing of climate change as a public health issue in US newspapers. *Environmental Communication, 10*(5), 593–611. doi:10.1080/17524032.2015.1050436

Wu, X., Zu, Y., Zhou, S., Chen, L., & Xu, B. (2016, January). Impact of climate change on human infectious diseases: Empirical evidence and human adaptation. *Environment International, 86,* 14–23.

4. *Online Coverage of Paris 2015: The View from the BBC*

NEIL T. GAVIN

Introduction

The COP21 negotiations in Paris 2015 were a significant milestone in the development of policies to address climate change. This was a 'critical discourse moment' (Carvalho & Burgess, 2005, p. 1461) where a high political profile and the associated media attention offer the opportunity for the (re) construction of public debate on the issues. In what follows, I assess the coverage on the BBC's news website around the COP21 conference in Paris in November–December 2015. I explore the sources citizens are most likely to trust and use as a source of information and news, arguing that we still need to focus on the broadcasters, and on their web output. The subsequent assessment focuses on the BBC News website stories during COP21, and—where appropriate—compares this with BBC terrestrial television coverage.

The Context for Assessing Web Coverage

There is much discussion about the digital and social media revolutions, and their implications for the way politics and public debates are conducted. However, even in this environment, it is as well to note that conventional media have a strong web profile, and dedicate considerable energy in developing their digital output, and promoting it through social media (Gavin, 2017). And there is a vast wealth of current evidence that the 'legacy' media—particularly the broadcasters—still have an important presence, and have a significant role in civic discourse, both in general terms, but also with regard to climate science and climate change.

The evidence suggests that, in the UK at least, television is far ahead of all other media—particularly the web and social media—as a news source for the public (Ofcom, 2007, 2013a). It is also reported by survey respondents as being ahead of newspapers and the web, as a news source for particular topics: science (Ipsos-MORI, 2014), climate change (Anderson, 2011; Painter, 2014; Whitmarsh, 2009), political and government issues (Newman, 2015; Newman & Levy, 2014), and elections.[1] Furthermore, the general public (YouGov, 2012), as well as elites (Populus, 2012), trust or prefer television over other sources, again by a substantial margin.[2] And broadcasters—particularly the BBC—are viewed as much less biased and more balanced than the press (YouGov, 2016) and, by an enormous margin, one of the most prominent online news aggregators (BBC, 2013). Now it is important to register that television news is, as a news source, in some decline (as it also is in the US), while the use of the internet is on the rise, particularly (but not exclusively) among the 16–24 age group (Ofcom, 2015, p. 7). However, the overall political significance of this is somewhat moderated by the fact that the younger age groups are least likely to be engaged in politics or to vote.

Nevertheless, it is evident that relatively few of those most likely to use social media, are doing so motivated by the urge to share content from the legacy media (Deloitte, 2015, p. 15). But we also know that the social media often refract content from their conventional counterparts, and that the public often refer their social contacts to the BBC News website (Newman, Dutton, & Blank, 2012, pp. 11–12). Furthermore, of those people who use their PC or mobile applications as news sources—rather than television or press—by far the biggest proportion access the BBC website and its associated apps (Ofcom, 2013b, p. 48). This, and the public's trust in broadcast output in particular, might go some way to explaining why the BBC website has a lot more traffic than all its conventional and new media equivalents.[3]

Research on climate change coverage is more often conducted on the output from the press than from television (Schäfer & Schlichting, 2014). In the absence of an analogue to full-text databases of content—like Lexis-Nexis—assessments of television's climate change coverage is difficult, and therefore not that common (for an overview, see Gavin, 2017). There are only two broad-ranging and detailed studies of UK television coverage around high-profile scientific reports or COP conferences, such as Copenhagen 2009 or Paris 2015. But although Gavin and Marshall (2011a) encompass 'the web' via Google searches, and O'Neill, Williams, Kurz, Wiersma, and Boykoff (2015) look at Twitter, neither explored the content from a source the public is known to trust and source—the BBC News website. There is a fair amount of research on BBC web output (for example, amongst others, Schifferes &

Coulter, 2013; Wahl-Jorgensen, Sambrook, & Berry, 2013), but little dealing with 'science' (Mellor, Webster, & Bell, 2011) or climate change (Holmes, 2009). What follows is such an assessment, focusing on BBC web-based coverage during the Paris COP21 negotiations in 2015.

The BBC Website During Paris COP21

The structure of the BBC News website in which such coverage is embedded is fairly stable, and has three principal areas.[4] First, at the top of the page are around fourteen bold, attention-grading, 'headline' stories. This section, occasionally carries multiple auxiliary reports—including older archived items—stacked in hyperlinks under the headlines, if a story is sufficiently big. Second, and further down the page, is a much larger block of stories and items. These are on the rest of the 'front page', and are made up of a wide variety of material: some in-depth stories, sport and regional items, weather and market/shares/currency details, some short video clips, and a picture gallery. At the bottom—but also buried under tabs at the page top—are the lead stories in each of eight specialist sections: business, UK politics, health, education & family, technology, entertainment & 'magazine' (a lifestyles/miscellany section).

The focus for what follows are stories in this space, in particular those featuring 'climate change' and 'COP' (or very close analogues) in their title and/or first four paragraphs. The start and end dates for analysis were chosen on two grounds. First, there is circumstantial evidence that conventional media coverage precedes and prompts attention to the web (Gavin, 2010, pp. 465–466). BBC television certainly gave COP21 considerable attention during the Paris conference (see Table 4.1). Second, the BBC web coverage showed a comparable distribution in roughly the same time frame (see Table 4.2). So our analysis focuses on the period when sustained broadcast coverage started and declined.

Table 4.1: BBC and ITN late/early evening bulletin containing climate change packages.

	28th November to 14th December																
Date	28	29	**30**	1	2	3	4	5	6	7	8	9	10	11	**12**	13	14
BBC early evening		*	*	*			*			*		*	*	*	*	*	
BBC late evening		*	*	*			*			*	*	*	*	*	*	*	

Note: The asterisks denote the date on which a single television bulletin package was broadcasts, while blanks denote no coverage; the dates in bold are when the Paris COP21 conference began and ended.

Table 4.2: BBC web pages featuring climate change.

	28th November to 14th December																
Date	28	29	**30**	1	2	3	4	5	6	7	8	9	10	11	**12**	13	14
Headlines	0	9	8	1	0	0	0	1	0	1	0	0	1	1	8	2	0
Front page	1	2	3	4	1	0	1	1	1	1	2	1	5	5	1	2	2
Science & environment	9	14	16	10	19	12	10	6	12	11	14	18	11	13	12	13	10

Note: The dates in bold are when the Paris COP21 conference began and ended.

Table 4.2 suggests there was a prodigious quantity of material offered the public (275 items over a seventeen day period). This is certainly congruent with the idea that the web offers more space to elaborate on issues than could be offered in the limited number of standard 2–3 minute slots that science and climate change stories are generally given (Hargreaves, Lewis, & Speers, 2003). However, the number of stories catalogued in Table 4.2 is somewhat deceptive, as there was a great deal of reuse of existing reports, rather than the constant creation of wholly new ones. The actual number of *unique* items is described in Table 4.3. When duplicated stories are excluded, there were eighty six of these items in total. But the stories they carried only hit the 'headline' slot in any numbers around the start and, to a lesser extent, the end of the conference. And in period assessed, climate change did not make the headlines at all on nearly half of the days examined.

Table 4.3: Unique BBC web pages featuring climate change.

	28th November to 14th December																
Date	28	29	**30**	1	2	3	4	5	6	7	8	9	10	11	**12**	13	14
Headlines	0	8	4	1	0	0	0	1	0	1	0	0	1	1	3	2	0
Front page	1	1	0	4	1	0	1	0	1	1	2	1	1	1	0	2	0
Science & environment	4	3	3	2	3	4	1	1	1	2	7	6	3	3	0	4	0

Note: The dates in bold are when the Paris COP21 conference began and ended.

How, in this context, should we approach the contribution to public debate of this body of reporting? Duplication aside, there was much for the public to peruse, should they choose to do so. However, it has been noted elsewhere

(Gavin & Marshall, 2011a; Jansen & Spink, 2005, 2006) that the public are highly unlikely to cruise through prodigiously quantities of web output or the associated hypertext connections on any given website, and rarely advance much beyond the first page of content. So the focus for what follows is the headline and 'front page' content outlined in Table 4.3, thirty nine stories in total.

Assessing BBC Web Coverage

One issue that can be addressed beyond a surface comparison of number of stories is the depth of commentary offered in them. Unlike the average broadcast package, the web gives almost unlimited space to elaborate on the topics addressed. The twenty one terrestrial TV packages highlighted in Table 4.1 had an average of just under six hundred words of news commentary. However, our assessment shows that this was almost exactly the same as the average for the thirty nine web stories examined. Nevertheless, although story lengths are comparable on web and TV, there were nearly twice as many of the former than the latter.

But did these more numerous BBC web stories cover a wider range of sub-issues? Like the television packages, each of the web stories tended to revolve around a single story topic, easily grouped into clusters of items covering quite similar themes. Table 4.4 outlines the result of this exercise, where the main theme of each story was categorised. By far the largest group of items dealt, not unexpectedly, with the Paris negotiations: their development, associated diplomatic difficulties, the details of what outcomes were sought, and which eventually materialised. The second largest category of items—ten in all—were dedicated, in some detail, to specific policy areas. Three of these related specifically to the UK. One flagged carbon capture techniques and the collapse of government policy to develop trial facilities ('Will carbon capture

Table 4.4: BBC web page news topics.

	N	%
Conference procedures, problems and developments	15	38.4
Specific policies	10	25.6
Climate protests	6	15.3
Climate change impacts	6	15.3
Climate science	2	0.5

ever happen in the UK?'). A second started with the same theme, but went on to look in detail at UK investment in mitigation and adaptation initiates abroad, and controversy over whether the associated funds were diverted from the existing foreign aid budgets ('COP21: UK under fire on climate policy'). A third story looked at the portfolio of UK climate-related policies, defended by Energy Secretary Amber Rudd, and balanced against some observations and criticisms from a range of political, NGO, and business commentators ('COP21: UK "will deliver" on climate change, says Amber Rudd').

However, the bulk of 'policy' stories were not about Britain. These were diverse in nature. Some covered generic issues, like the politics of meat consumption ('Can eating less meat help reduce climate change?', and 'Arnold Schwarzenegger: "Go part-time vegetarian to protect the planet"'), forestry and deforestation ('COP21: Business needs to act on forests, says Charles') and renewable energy ('Attenborough: "Solar energy should undercut oil"'). Where they were specifically located, they dealt with the recycling of human waste in Ghana ('COP21: Ghana's green energy gamble on "poo-power"'); coal-fired energy production and policy in China, India, Indonesia, and the EU ('COP21: Coal plans would derail 2 degree warming target'); and an inventory of snapshots on policy initiative by the mayors in a range of cosmopolitan cities ('COP21: City mayors discuss green solutions').

A smaller number of stories dealt with climate protests in Paris. Unlike Copenhagen in 2009 (Gavin & Marshall, 2011b), these were reported as uniformly peaceful in nature. However, one dealt with an element of mischievousness bordering on criminality, where there was the unauthorised replacement of Paris street poster advertising, with climate messages ('COP21: Eco activists Brandalism launch Paris ad takeover'). Finally, for a conference premised on what we know about climate change, only two stories were specifically about, and detailing, climate science topics: one on oceanic acidification ('COP21: What does ocean acidification look like?'), and another on current and projected world temperatures, CO_2 emissions and its top ten producers, and Arctic ice retreat ('Six graphics that explain climate change'). However, the science was implicit in a larger group of stories dealing with the *impacts* of climate change. Here only one touched on Britain ('Five ways climate change could affect the UK'). The rest dealt with the retreat of the Rhone glacier in Switzerland, an item on typhoons in the Philippines, one on climate-induced landslides in Alaska, and one apiece on sea level rises in Tonga and Vanuatu.

Was this different from the picture painted by the conventional BBC broadcasts outlined in Table 4.1? At one level the answer is no, since the broad themes in stories on television were easily accommodated in categories set out in Table 4.4. Table 4.5[5] shows that the largest single category dealt

Table 4.5: BBC broadcast news topics.

	N	%
Conference Procedures, Problems and Developments	11	42.3
Specific Policies	5	19.2
Climate Protests	2	7.6
Climate Change Impacts	7	26.9
Climate Science	1	3.8

with the COP21 negotiations. However, there were a greater proportion of stories on broadcast TV about the science and, particularly, climate change impacts, than there were on the web. There was only one story categorised under 'climate science', dealing with the various carbon outputs from British homes (early evening, 10 December). However, the impact on British retail's supply chains was covered in one item (early evening, 4 December), and two dealt with actual or likely flooding effects (early and late evening, 7 December). Two further stories mentioned, respectively, the Cook Islands and China (late evening, 30 November) and the Pacific Islands, Brazil, Bangladesh, and the Philippines (late evening, 11 December).

On specific policy issues, there were two virtually identical packages (early and late evening, 13 December) outlining UK government plans to close coal fired power plants, alongside some criticism of subsidy curbs for renewables, shale gas expansion and continued oil and gas use. One other UK oriented item saw the Unilever CEO criticise the government's removal of financial support for wind and solar power. Beyond the UK, one story outlined political controversy in the US over government policy on reducing power plant emissions (late evening, 1 December). Another report dealt with general underfunding of renewables, alongside China's investment in this technology, while flagging its reliance on coal as an energy source (late evening, 30 November). Finally, only two reports dealt with protest, both briefly flagging some clashes with the police (early and late, 29 November), though emphasising that most were peaceful and quiet, even 'solemn'.

The picture painted here suggests a rich *overall* communication environment. The web and terrestrial output from the BBC has some overlap, but the complementary sets of stories cover a kaleidoscope of topics and sub-themes, dispersed over a fairly wide range of locales. There might have been more on both media about the science, since there is still considerable public confusion about whether the climate is changing, and the role of human activity in the process.[6] Nevertheless, the emphasis on climate change impacts beyond Europe is consistent with scientists' perceptions of which locales will be most adversely affected, and when (Bray & von Storch, 2010).

Was anything significant missing from the web-based stories the public were most likely to see during the Paris conference? The answer to this is a provisional yes. For instance, in none of the web stories surveyed (or, indeed, in the television coverage) were aviation and shipping policies the principal story themes. This is despite the fact these sectors of the economy make a very substantial contribution to world CO_2 emissions, have been controversial in negotiations at both Copenhagen and Paris, and—perhaps more importantly—are the veins and arteries of contemporary globalised trade. By the same token, there was nothing on market failures, or the 'externalities' that Lord Stern alluded to when he noted that 'Climate change is a result of the greatest market failure the world has seen'. And finally, there was almost no attention to the role of consumer-and-consumption-based activities in the creation of the global warming problem, save a single quote from an activist whose stated aim was to '… highlight the links between advertising, consumerism, fossil fuel dependency and climate change' ('COP21: Eco activists Brandalism launch Paris ad takeover').

Voices on the BBC

Did the BBC web and broadcast reports offer a platform for a wide range of perspectives and positions? One way to assess this is with reference to the individuals allowed to give voice to their opinions in the commentary. Table 4.6a and 4.6b presents this in detail, scooping speakers into nine categories: 'vox pop', academics, NGOs and think tanks, United Nations officials, business representatives, and a catchall 'others'. There were also three categories of politician: UK national, Europe and EU, and international, though it should be acknowledged that these are not necessarily exclusive groupings.

Table 4.6a illustrates the actors quoted in web stories, while Table 4.6b shows those present in BBC television coverage. There are some unsurprising dimensions to Table 4.6. For instance, by far the largest grouping are politicians, who were the principal players in the drama that unfolded in Paris. The emphasis on them is consistent with known journalistic practices and norms, which favour 'authority figures' (Boykoff & Boykoff, 2007), politically significant themes (Hetherington, 1985) and elites, especially in 'elite nations' (Harcup & O'Neill, 2001).

However, the distribution in the groups across the web-television divide were not always as expected. The web and television stories had roughly

Table 4.6a: Voices quoted in web-based stories.

Vox Pop	**Academic**	**NGO/Think Tank**	**Politicians (International)**	**Politicians (UK National)**	**Politicians (Europe and EU)**
Kate Charlesworth, a doctor and mother, Sydney	Professor Corinne Le Quere, University of East Anglia (**2**)	Craig Bennet, Chief Executive of Friends of the Earth (**3**)	Todd Stern, US lead negotiator (**2**)	David Cameron, PM (**2**)	Laurent Fabius, French Foreign Minister (**5**)
Vincent Morrish, manager of Tongan hotel	Dr David Volken, Swiss Federal Office of the Environment	Helen Szoke, Oxfam (**2**)	Prakash Javadekar, Environment Minister, India (**2**)	Dr Aileen McLeod, Scottish Parliament Minister for the Environment	President Francois Hollande (**3**)
John Kalmatak, Vanuatu islander	Professor Stuart Haszeldine, Director, Scottish Carbon Capture and Storage	Stop Climate Chaos Scotland	Antolin Huascar, a Peruvian agricultural official	Patrick Harvie MSP, co-convener of the Scottish Greens	Miguel Arias Canete, EU Commissioner
'Marinel', Philippine typhoon victim	Sione Fulivai, senior climate finance analyst, Tonga	Tom Ballantine, Chairman of Stop Climate Chaos Scotland	Amjad Abdulla, Chair of the Alliance of Small Island States	Sarah Boyack, Scottish Labour's environmental justice spokesperson	Laurence Tubiana, French climate ambassador
	Professor Jason Hall-Spencer, University of Plymouth	Tom Burke, E3G	Vladmir Putin, Russian President	Jeremy Corbyn, leader of the Labour Party	Miguel Arias Canete, EU's Climate Commissioner
	Dr Margaret Darrow, University of Alaska	Siale 'Ilolahia, Director, Civil Society Forum of Tonga'	Enele Sopoaga, Tuvalu PM	Boris Johnson, London Mayor	

(*Continued*)

Table 4.6a: (*Continued*)

Vox Pop	Academic	NGO/Think Tank	Politicians (International)	Politicians (UK National)	Politicians (Europe and EU)
	Dr Ronald Daanen, Alaska Department of Natural Resources	Dr Niklas Hohne, Climate Analytics, Berlin	Eduardo Paes, Rio Mayor	Darren Johnson, Green member for London Assembly	
	Professor Joanna Haigh, Grantham Institute, Imperial College London	Bill Hare, Climate Analytics, Berlin	Frank Jensen, Copenhagen Mayor	Unnamed Department of Environment and Climate Change spokesperson	
	Professor Paul Bates, Royal Society	Markus Hagemann, NewClimate Institute	Clover Moore, Sydney Mayor	Michael Jacobs, climate adviser to UK PM Gordon Brown	
		Tasneem Essop, WWF	Alfred Okoe Vanderpuije, Accra Mayor	Liz Truss, Environment Secretary	
		David Waskow, World Resources Institute	Karin Wanngård, Stockholm Mayor	Amber Rudd, Energy Secretary	
		Michael Jacobs, New Climate Economy	Gregor Robertson, Vancouver Mayor	Nicola Sturgeon, Scottish First Minister	

		Jennifer Morgan, World Resources Institute	Nozipho Mxaka-to-Diseko, South Africa climate ambassador	Barry Gardiner, Labour Shadow Energy Secretary	
		Adriano Campolina, ActionAid	Thoriq Ibrahim, Chair of the Alliance of Small Island States	Richard Benyon, former Environment Minister	
		David Nussbaum, WWF-UK Chief Executive	Unnamed 'Least Developed Countries' spokesperson		
		Nick Deaden, Global Justice Now	Gurdial Singh Nijar, 'Like Minded Group' spokesperson		
United Nations	Business	Others	Arnold Schwarzenegger, former California Governor		
Ban Ki-Moon, UN Secretary General (**2**)	Carolyn Fairbairn, CBI Director-General	Prince Charles (**2**)	Tony De Brum, Marshall Islands Foreign Minister		

(*Continued*)

Table 4.6a: (*Continued*)

Vox Pop	Academic	NGO/Think Tank	Politicians (International)	Politicians (UK National)	Politicians (Europe and EU)
Christiana Figueres, Head UN climate change negotiations	World Business Council for Sustainable Development spokesperson	Dame Vivienne Westwood, designer	John Kerry, US Secretary of State		
Unnamed UN climate spokesperson		Peter Gabriel, musician	Peter Woolcott, Australian environment ambassador		
		Sir Richard Attenborough, BBC presenter	Jo Tyndall, New Zealand Climate Change Ambassador		
		Source 'close to group of poorer countries'	Barack Obama, US President		
			Xie Zhenhua, China's chief negotiator		
			Narendra Modi, India PM		
			Giza Gaspar Martins, Chairman 'Least Developed Countries'		

Note: Where the same individual appeared in more than one separate story, the number of their appearances is added in parenthesis.

Table 4.6b: Voices quoted in broadcast stories.

Vox Pop	**Academic**	**NGO/Think Tank**	**Politicians (International)**	**Politicians (UK National)**	**Politicians (Europe and EU)**
Tuaine Marsters, Cook Islands resident (2)	Professor Dame Julia Slingo, Chief Scientist, Met Office (2)	Samantha Smith, WWF	Barack Obama, US President (2)	Amber Rudd, Secretary of State for Energy and Climate Change (2)	President Francois Hollande (2)
Antonia Martinez, Spanish farmer	Professor Kevin Anderson, Tyndell Centre for Climate Change Research (2)	Tom Steyer, President, NextGen Climate	John Kerry, US Secretary of State (2)	Lisa Pinney UK Environment Agency	Laurent Fabius, French Foreign Minister
Andrew Lingham, UK farmer	Professor Tim Jackson, University of Surrey		Prakash Javadekar, Environment Minister, India (2)		Carole Dieschbough, European Union Representative
Unnamed male UK flood victim	Ben Caldecott, Smith School, University of Oxford		Narendra Modi, India PM		
Unnamed female UK flood victim	Lord Stern, Grantham Research Institute		Denis Lowe, Caribbean Representative		
Isobel McLauchlin, UK householder	Professor Corinne Le Quere, Tyndall Centre for Climate Research		Anote Tong, Kiribati President		

(*Continued*)

Table 4.6b: (*Continued*)

Vox Pop	**Academic**	**NGO/Think Tank**	**Politicians (International)**	**Politicians (UK National)**	**Politicians (Europe and EU)**
Qie Qun Jia, Tibetan herdsman			Gao Feng, Chinese Climate Representative		
Shen Youguo, Chinese solar engineer			Tony La Vina, Philippines Spokesman		
Scott Gediman, Park Ranger, Yosemite National Park			Nozipho Mxakato-Diseko, South African Representative		
Unnamed US miner			Dr Paul Oquist Kelley, Nicaraguan Representative		
'Monica' Namibian farmer					
Unnamed Namibian					
UNITED NATIONS	BUSINESS	OTHERS			

Ban Ki-Moon, UN Secretary General (**4**)	Bill Gates, Micro-soft (**2**)	Emma Thompson, actress (**2**)			
Victoria Tauli-Cor-puz, UN Human Rights official (**2**)	Andy Clark, Chief Executive, Asda	Sir Richard Atten-borough, BBC presenter			
	Lyndon Rive US Chief Executive, SolarCity	Prince Charles			
	Bill Raney, Presi-dent, West Virginia Coal Association	Selina Leem, Youth Representative, Marshall Islands			
	Paul Polman, Chief Executive of Unilever				
	Juliet Devenport, Good Energy				

Note: Where the same individual appeared in more than one separate story, the number of their appearances is added in parenthesis.

comparable sizes, in terms of word length. But there were twice as many of the former than the latter. We might, therefore, expect that, for each category of actor, there would be greater capacity to showcase a larger number of voices on the web. This was certainly the case for academics, international and European politicians, and UN actors. UK national politicians, NGOs, and think tanks do, however, seem to be somewhat over-represented on the web, and under-represented on television. And clearly, a wider variety of 'vox pop' and business voices were offered more space on television than on the web, with no obvious journalistic norms consistently at play.

Regardless of differences in the weighting of various categories of actor across web and television commentary, the commanding feature of Tables 4.6a and 4.6b is the wide variety of voices that the BBC web output adds to the broader communication environment. In just over two weeks of coverage there were just under one hundred and sixty quoted sources across BBC television and web output. Of these, nearly two thirds were represented on the web space. These were from a very wide range of countries, and not just the dominant 'elite nations' journalists are said to favour (Harcup & O'Neill, 2001). And the actors engaged were also from a plethora of nationally and internationally active third sector organisations, and from across the political divide.

One final, but important, point that emerges from Tables 4.6a and 4.6b. None of the actors quoted were known climate sceptics. These actors have managed to insinuate themselves into BBC coverage during the Copenhagen negotiations (Gavin & Marshall, 2011a) and have surfaced in BBC television commentary around the UN IPPC science reports in 2013–2014 (Gavin, 2016). And they have consistently surfaced in newspaper reports (Painter, 2010; Painter & Gavin, 2016). However, they were given no platform in either BBC television or web commentary during the Paris negotiations.

Conclusions

The output from broadcasters is still a very significant—perhaps still *the* most significant—source of information and news for the British public. The BBC's web-based coverage may not be the first place citizens go to get an appreciation of what is going on. But it is, nevertheless, a source that is used, and which is likely to be trusted. The assessment offered in this chapter suggests some provisional conclusion that can be drawn from an examination of the content in this source.

Evidently the BBC web space was used to extend, to a very considerable degree, the amount of coverage the organisation was able to offer via the conventional terrestrial broadcasting route. We did not assess all of this prodigious web output, but only those parts the public were most likely to see. We found that the BBC's coverage of protests was not as potentially damaging to the protesters' image as it was at the Copenhagen negotiations in 2009 (Gavin & Marshall, 2011b). And, in terms of the broader themes addressed and the size of the stories about them, the web coverage was not too dissimilar to conventional television output. But it did however extend, by some way, the range of sub-themes addressed across the BBC's spectrum of output.

Moreover, and importantly, the web reports offered space for a great many institutions, organisations, individuals, and political actors to have a say. And there was not an obvious or commanding '… preference for official and elite sources'. (Wahl-Jorgensen, Berry, Garcia-Blanco, Bennett, & Cable, forthcoming, p. 5) in sourcing, that others have detected, especially with respect to academics, developing country representatives, and NGO or think tank actors. This is a positive contribution to public engagement with these organisations and actors, as well as public engagement with the messages they voice. A further contribution to public understanding might also be seen in the absence of climate sceptic voices. Competing climate science and climate sceptic voices are known to confuse and frustrate the public (Butler & Pigeon, 2009; Shuckburgh, Robison, & Pidgeon, 2012). But none such were in evidence on the BBC's web commentary, though an assessment of the details of the reports would be required to determine if messages paralleling sceptical themes were buried in the commentary.

Finally, it should be noted that some important issues were perhaps under-reported in the web commentary that the public were most likely to see around the Paris COP21 conference—notably aviation, shipping, consumerism and market failure. This also has a bearing on public understanding, especially where their contribution to the climate change threat is well understood and, for the most part, commonly acknowledged. So there were gaps in the coverage. And again, an analysis of story details would be necessary to judge whether these absences extended, for instance, to other cause of climate change, or to themes missing from reports about the impacts of climate change.

Acknowledgement

The author wishes to acknowledge the efforts of volunteer research assistance by students of Politics Department, Liverpool University, in the creation of the datasets used in this study, particularly the work of Rachel Harrison and Matthew Ward.

Notes

1. http://www.campaignlive.co.uk/article/1344407/election-2015-people-news
2. http://www.on-broadcast.com/blog-category-1/we-asked-100-mps-whats-your-favourite-media/
3. https://www.similarweb.com/blog/index-top-u-k-media-publishers-and-publications-of-2015
4. http://www.bbc.co.uk/news
5. Four of these stories were longer and had considerable copy on more than one core theme. These themes have, therefore, been counted independently.
6. https://yougov.co.uk/news/2013/04/03/climate-change/

References

Anderson, A. (2011). Sources, media, and modes of climate change communication: The role of celebrities. *Wiley Interdisciplinary Reviews: Climate Change, 2*(4), 535–546.

BBC. (2013). *Public perceptions of the impartiality and trustworthiness of the BBC.* London: BBC.

Boykoff, M. T., & Boykoff, J. M. (2007). Climate change and journalistic norms: A case-study of US mass-media coverage. *Geoforum, 38*(6), 1190–1204.

Bray, D., & von Storch, H. (2010). *CliSci2008: A survey of the perspectives of climate scientists concerning climate science and climate change.* Geesthacht: GKSS-Forschungszentrum Geesthacht.

Butler, C., & Pigeon, N. (2009). Media communications and public understanding of climate change: Reporting scientific consensus on anthropogenic warming. In T. Boyce & J. Lewis (Eds.), *Climate change and the media.* Oxford: Peter Lang, 43–58.

Carvalho, A., & Burgess, J. (2005). Cultural circuits of climate change in UK broadsheet newspapers, 1985–2003. *Risk Analysis, 25*(6), 1457–1469.

Deloitte. (2015). *Media consumer 2015: The signal and the noise.* London: Deloitte.

Gavin, N. T. (2010). Pressure group direct action on climate change: The role of the media and the web in Britain—A case study. *British Journal of Politics and International Relations, 12*(3), 459–475.

Gavin, N. T. (2016). Media, environment and global security. In P. Robinson, P. Seib, & R. Fröhlich (Eds.), *Routledge handbook of media, conflict and security.* London: Routledge. 295–308.

Gavin, N. T. (2017). TV and cable news coverage of climate change. In M. Nisbet, S. O'Neill, E. Markovitz, S. Ho, J. Thaker, & M. S. Schäfer (Eds.), *Oxford research encyclopaedia of climate science.* Oxford: Oxford University Press. DOI: 10.1093/acrefore/9780190228620.013.360

Gavin, N. T., & Marshall, T. (2011a). Mediated climate change in Britain: Scepticism on the web and on television around Copenhagen. *Global Environmental Change, 21*(3), 1035–1044.

Gavin, N. T., & Marshall, T. (2011b). Climate change and international protest at Copenhagen: Reflections on British television and the web. In S. Cottle & L. Lester (Eds.), *Transnational protests and the media*. Oxford: Peter Lang. 197–212.

Harcup, T., & O'Neill, D. (2001). What is news: Galtung and Ruge revisited. *Journalism Studies, 2*(2), 261–280.

Hargreaves, I., Lewis, J., & Speers, T. (2003). *Towards a better map: Science, the public and the media*. Swindon: Economic and Social Research Council.

Hetherington, A. (1985). *News, newspapers and television*. London: Macmillan.

Holmes, T. (2009). Balancing act: PR, 'impartiality' and power in mass media coverage of climate change. In T. Boyce & J. Lewis (Eds.), *Climate change and the media*. Oxford: Peter Lang. 92–100.

Ipsos-MORI. (2014). *Public attitudes to science*. London: Department for Business, Innovation and Skills.

Jansen, B. J., & Spink, A. (2005). An analysis of web searching by European AllltheWeb.com users, *Information Processing Management, 41*(2), 361–381.

Jansen, B. J., & Spink, A. (2006). How are we searching the world wide web? A comparison of nine search engine transaction logs. *Information Processing Management, 42*(1), 248–263.

Mellor, F., Webster, S., & Bell, A. (2011). *Content analysis of the BBC's science coverage: Review of the impartiality and accuracy of the BBC's coverage of science*. London: BBC.

Newman, N. (2015). *Reuters institute digital news report 2015*. Oxford: Reuters Institute for the Study of Journalism.

Newman, N., Dutton, W. H., & Blank, G. (2012). Social media in the changing ecology of news: The fourth and fifth estates in Britain. *International Journal of Internet Science, 7*(1), 6–22.

Newman, N., & Levy, D. A. L. (2014). *Reuters institute digital news report 2014*. Oxford: Reuters Institute for the Study of Journalism.Ofcom. (2007). *New news, future news*. London: Author.

Ofcom. (2013a). *News Consumption in the UK—2013 report*. London: Author.

Ofcom. (2013b). *News Consumption in the UK—2013 slides*. London: Author.

Ofcom. (2015). *News consumption in the UK: Research report*. London: Author.

O'Neill, S., Williams, H. T. P., Kurz, T., Wiersma, B., & Boykoff, M. (2015). Dominant frames in legacy and social media coverage of the IPCC fifth assessment report. *Nature Climate Change, 5*(4), 380–385.

Painter, J. (2010) *Summoned by science: Reporting climate change at Copenhagen and beyond*. Oxford: Reuters Institute for the Study of Journalism.

Painter, J. (2014) *Disaster averted? Television coverage of the 2013/14 IPCC's climate change reports*. Oxford: Reuters Institute for the Study of Journalism.

Painter, J., & Gavin, N. T. (2016). Climate skepticism in British newspapers, 2007–2011. *Environmental Communication, 10*(4), 432–452.

Populus. (2012). *Power, principles and the press*. London: Author.

Schäfer, M., & Schlichting I. (2014). Media representations of climate change: A meta-analysis of the research field. *Environmental Communication, 8*(2), 142–160.

Schifferes, S., & Coulter, S. (2013). Downloading disaster: BBC news online coverage of the global financial crisis. *Journalism, 14*(2), 228–252.

Shuckburgh, E., Robison, R., & Pidgeon, N. (2012). *Climate science, the public and the news media: Summary findings of a survey and focus groups conducted in the UK in March 2011.* London: NERC.

Wahl-Jorgensen, K., Berry, M., Garcia-Blanco, I., Bennett, L., & Cable, J. (2018) Rethinking balance and impartiality in journalism? How the BBC attempted and failed to change the paradigm. *Journalism.*

Wahl-Jorgensen, K., Sambrook, R., & Berry, M. (2013) *BBC breadth of opinion review: Content analysis.* Cardiff: Cardiff University.

Whitmarsh, L. (2009). What's in a name? Commonalities and differences in public understanding of 'climate change' and 'global warming'. *Public Understanding of Science, 18*(4), 401–420.

YouGov. (2012). *Whom do the public trust?* London: Author.

YouGov. (2016). *38 Degree survey.* London: Author.

5. *Climate Change in Brazilian Media*

Anabela Carvalho and Eloisa Beling Loose

Introduction

Brazil's contribution to global greenhouse gas emissions (mainly via deforestation), the importance of the Amazon as a carbon sink and 'controlling agent' of several meteorological processes, and, according to global surveys, the population's unusually high awareness of climate change make the country of great interest in the study of climate change communication. Brazil is also home to some of the world's largest stocks of biodiversity and of fresh water, as well as a major hydropower and ethanol producer (which have their own—in some cases massive—environmental impacts).

Given a widespread dependence on natural resources and profound social inequalities, Latin America is quite vulnerable to the impacts of climate change. In Brazil, this is aggravated by issues such as demographic pressure, disorganized urban sprawl, and poor governance situation (Painel Brasileiro de Mudanças Climáticas, PBMC, 2013). The country has undergone rapid transformations, with large-scale economic and social improvements in the first decade of the twenty-first century followed by a significant decline in the last few years. Politically, the country's fragile democracy has also seen recent dramatic challenges.

In these complex contexts, analysing the ways in which Brazilian media construct and revise the meanings of climate change is of crucial importance. This chapter will start by providing contextual information on Brazil, then examine public perceptions of climate change and the Brazilian media system, discuss research on media coverage, and finally, identify gaps and future research opportunities.

Climate Change and Climate Politics in Brazil

A significant contribution to the greenhouse effect, severe (projected) impacts, high vulnerability, and erratic policies (especially in the last few years) are key traits of Brazil's climate-related situation, as we will discuss below.

It is well known that most developing countries are badly exposed to the consequences of climate change (IPCC, 2014). They are likely to bear the brunt of climate impacts and the vast majority of the cost of associated loss and damage. Forecasts for Latin American point, inter alia, to the disappearance of the Andean glaciers, a crucial source of water (and hydropower generation) for many millions of people, and the destruction of the Amazon forest and subsequent transformation into a savannah-type landscape, which would lead to further changes in the region's climate and possibly the whole world's. With limited technical and financial capacity to respond to climate change effects, most Latin American countries have a low degree of resilience (Vergara et al., 2013).

Due to its continental dimensions, Brazil has a great diversity of climatic regimes and influences on its climate, which hinders the acquisition of series of observational data in extended periods. Existing data indicates a significant reduction of rainfall in much of the Centre, North, and Northeast of Brazil, whereas in the South and Southeast episodes of very heavy rainfall have increased in the last 50 years. Climatic models suggest that the frequency of occurrence of extreme and prolonged droughts will rise in the future (especially in the Amazon, Cerrado and Caatinga biomes). Projections also point to a warmer future due both to greenhouse gas emissions and to regional processes such as urbanization and deforestation (PBMC, 2013).

While forecasts point to severe impacts of climate change across the country, a number of factors increase vulnerability. According to the Brazilian Panel for Climate Change (PBMC, 2013) demographic pressure, disorderly urban growth, poverty and rural migration, low investment in infrastructure and services, and governance issues all contribute to increase the country's vulnerability. High concentration of population in urban areas puts most Brazilian cities at further risk in relation to the effects of climate change (PBMC, 2013).

Brazil is the world's seventh highest emitter of greenhouse gas emissions (Friedrich, De, & Damassa, 2015), mostly due to deforestation. In the last few decades, a number of measures have been put in place to check for forest loss in the Amazon, such as satellite monitoring. Environmental legislation adopted further to the 1992 United Nations Conference on Environment and Development (UNCED; also known as Earth Summit or Rio 92) is progressive, although its implementation leaves much to be desired. In the year

2000, the National Climate Change Forum was created with the aim of raising awareness and public engagement. Examining the period 1989–2003, Viola (2004, n.p.) noted that Brazil moved from a 'more nationalist toward a more liberal and globalist standing' in international climate politics. In subsequent years it often played a leading role and acted as a bridge between developed and developing countries in international negotiations.

The National Policy on Climate Change (Política Nacional sobre Mudança do Clima) and the Brazilian Panel on Climate Change (PMBC) were established in 2009 and the government made a voluntary commitment to reduce deforestation by 80% until 2020. Policies in the domains of conservation and energy, amongst others, started taking climate change into account. Official efforts also include augmenting reforestation and protected areas, as well as enhancing efficiency in the cycle of ethanol (a widely used biofuel), developing seeds with more resistance to climate variability and technologies to capture methane in hydroelectric reservoirs. Research carried out by the Instituto Nacional de Pesquisas Espaciais (Inpe) and the Instituto de Pesquisas Amazônicas (Inpa) in studying and promoting climate change mitigation and adaptation were also positive developments. Until 2014, the country made remarkable progress in the reduction of deforestation, which was pushed by the action of non-governmental organizations and of political figures like Marina Silva with subsequent governmental commitment (Viola, 2013). For this, Brazil received praise as an international leader in reducing emissions (Howard, 2014).

Despite those notable developments, national climate politics has long been marred by conflicting policy decisions and significant delays in the execution of plans. Overall, much remains to be done. Mitigation potentials in the sectors of transport, energy-generation, industry, agriculture, and livestock remain underexplored. A chronic scarcity of funds to keep forests standing and reduce emissions has reduced intervention to market-based mechanisms, such as REDD (Reducing Emissions from Deforestation and Forest Degradation), whereby richer countries buy emissions credits from Brazil (Börner et al., 2010). Another weak point is the lack of regional and local policies for climate change in most Brazilian states, both in terms of mitigation and adaptation, which would be crucial to enact actions in the locales where greenhouse gas emissions are produced and where climate change impacts are being felt.

Economic recession and political turmoil in the last few years have cast further shadows on the country's role in climate politics. Conservation-oriented policies have been weakened, through, among other aspects, the loosening of the Forest Code (which regulates how Brazilian native vegetation can be explored) and cuts made in the areas of inspection and monitoring.

Despite signing the Paris Agreement, Brazil has adopted measures that are at odds with climate protection after it came into force, such as changes to regulations in oil exploration conducive to faster and larger extraction (Angelo, 2016). Recent records offer a dismal picture. According to the Greenhouse Gas Emissions Estimation System of Observatório do Clima (Climate Observatory), Brazilian greenhouse gas emissions had an annual increase of 3.5% in 2015 despite the country's GDP falling by 3.8% (Observatório do Clima, 2016). The rise in emissions was mainly due to the increase in deforestation, which grew 24% in 2015 compared to the previous year. The Instituto Nacional de Pesquisas Espaciais (2016), a federal government agency that runs a deforestation monitoring project, estimated a 29% annual increase in the rate of deforestation in 2016.

Public Perceptions of Climate Change

In a six-country study conducted by Gallup in 1992 (the year of Rio's Earth Summit), the Brazilian public stood out (together with the Portuguese) in considering global warming a very serious issue: 71% of respondents compared with 41% in Russia and 47% in the USA (Dunlap, 1998). By 2007, the Pew Research Center found that 88% of respondents in Brazil considered global warming a very serious problem, the highest percentage of all 47 countries polled in multiple world regions. That trend was observed again in 2015 (Pew Research Center, 2015) with 86% of Brazilians rating climate change as very serious, again the highest percentage among the polled countries. Brazilians (90%) were also 'the most likely to believe that climate change is harming people now' (p. 16). They expressed very high support for national action towards limiting emissions (88%) and were most likely to endorse 'modifications in the way they live to better cope with global warming' (89%) (p. 29).

In contrast with the Gallup and Pew surveys, surveys run by the Brazilian Environment Ministry (Ministério do Meio Ambiente, 2012) show that the environment was not perceived as one of Brazil's top ten problems in 1992, 1997, 2001, and 2006. It only came to that position in the 2012 survey (coincidently Rio+20 took place in 2012, giving rise to a new wave of media coverage of the environment). When asked, in 2012, to rank the world's most important environmental problems Brazilians answered in the following order: deforestation; pollution of rivers, lakes and other sources of water; air pollution; increasing waste; ozone layer; waste of water and energy in cities; ocean pollution; and, in eighth position, climate change, with only 10% of the answers (down from 43% in 2006). Climate change appeared in responses to this poll as one of Brazil's main environmental problems (6%

of answers) for the first time. Compared to the Gallup and Pew survey data, these results suggest that different questions lead to different images of the importance of climate change for the public. It is therefore important to ask and cross-compare diverse sets of questions. In any case, the Brazilian Environment Ministry data also indicates that there has been a rising awareness of and engagement with the environment, with significantly higher percentages of people expressing readiness to act on environmental issues in 2012 than in previous years.

The Brazilian Media System

The Brazilian media system is 'television-centered' (Albuquerque, 2012, p. 78). 77% of the population watches television daily (Secretaria da Comunicação Social da Presidência da República, 2016) and the small set occupies a key place in the country's culture and society. It is by far the most widely used source of information. 50% of Brazilians use the internet daily (a 15% increase in relation to 2015) and radio is the third most popular media (30% daily use). Brazilian newspapers have a very low rate of circulation (8% daily readership) and are 'addressed to a small urban elite' (Albuquerque, 2012, p. 78). Most of the press in Brazil is regional, although some newspapers are viewed as particularly influential at the national level, including *Folha de São Paulo, O Estado de S. Paulo* and *O Globo* (based in Rio de Janeiro, part of the Globo Group). Worth mentioning is also the news magazine *Veja*, with the highest sales rate, and others, such as *Isto É*, Época and *Carta Capital.*

Brazilian broadcasting media are (and have always been) almost exclusively private. Rede Globo is the biggest television network and part of the Globo Group, Latin America's largest media conglomerate. SBT, Record and Bandeirantes are other big players in the television business, which belong to multimedia groups owned by powerful families or—in the case of Record—an evangelical church.

Media oligopolization has significant political implications especially given the strong connections between corporate power and political power in Brazil (e.g. Fonseca, 2010). At the same time, the country also has a lively alternative media scene with multiple—but low scale—initiatives to produce independent information (Oliveira, 2011). However, a past record of authoritarian politics (without press freedom for long periods), various types of pressures and multiple insecurities, including physical attacks and murders, obstruct the exercise of journalism in Brazil, which Reporters without Borders have ranked in the 104th position in its 2016 World Press Freedom Index.

The Brazilian press has a predominant right wing orientation. Although most media have a 'market driven, catch-all attitude' and 'distance themselves from particular political groups' (Albuquerque 2012, p. 81), there are multiple signs that Brazilian media predominantly favour neoliberal views.

Media (Re)constructions of Climate Change

The emergence of environmental journalism in Brazil is tied to both national and international developments. Following international events and tendencies, coverage of environmental matters first rose in the Brazilian media in the early 1970s (Girardi, Moraes, & Loose, 2012). Costa (2006) found that attention to the environment and especially the Amazon intensified in the late 1980s in the main newspapers and news magazines. The dissemination of the first satellite images showing huge numbers of forest fires and the depletion of the Amazon forest, the assassination of rubber tapper and environmentalist Chico Mendes, and the emergence of climate change in international political and media arenas were among the factors fuelling media interest. Rio 92, the landmark UN summit where the United Nations Framework Convention on Climate Change was agreed and the Convention on Biological Diversity was signed, offered a critical opportunity for media analysis of environmental issues and further pushed them into news agendas. Research on environment and the media was relatively slow to develop and is not as extensive as in other countries/regions. Nonetheless, a number of works on the coverage of climate change published over the last decade or so allow for the identification of various trends.

Reflecting International Agendas and Foreign Voices

Various studies indicate that news coverage of climate change in Brazil is strongly influenced by international factors. In one of the first analyses, Fioravanti (2007) showed that between the second half of 2006 and the first half of 2007 the number of news stories in *Folha de São Paulo* increased four-fold, which he associated with events such as the launch of the various IPPC's Working Group reports and a G8 meeting. There was little focus on Brazil with attention being instead turned to consequences of climate change around the world. According to Fioravanti (2007), the newspaper climate change reporting displayed a strong dependency on international news agencies.

The broadest empirical analysis of Brazilian media coverage of climate change was carried out by ANDI, a civil society organization focusing on media, development, and young people's rights, which analysed

50 newspapers between 2005 and 2008 (Vivarta, 2010). Looking at coverage from July 2005 to June 2007 and July 2007 to December 2008, they identified fluctuations in media attention that they tied to international events. More specifically, they observed that the number of articles started rising at the end of 2006 with the publication of the Stern Report and the opening of Al Gore's documentary *An Inconvenient Truth,* and increased further in 2007 with the publication of the IPCC's Working Group reports and the awarding of the Nobel Peace Prize to Al Gore and the IPCC in 2007. Despite these general trends, the volume of references to national matters grew between the two (consecutive) periods that they analysed, which they attributed to a change in the Brazilian's government stance in relation to climate change and energy (from refusal to acceptance of emissions reduction commitments) (see also Broadbent et al., 2016).

Subsequent research corroborated that the Brazilian press tends to reproduce international agendas. Compared with newspapers from the USA, Argentina and Colombia, Zamith, Pinto, and Villar (2015) pointed to '*Folha de São Paulo* having the greatest proportion of international stories (69.9%)' on climate change in 2009. This tendency also appears to extend to more localized, less 'cosmopolitan' newspapers. Loose (2016) found a prevalence of foreign actors in reports of regional newspaper *Gazeta do Povo* (Paraná state) in 2013.

Privileging Official Sources

In his analysis of two leading newspapers' coverage of the Rio 92 summit, Reis (1999) found that more than 50% of sources cited were either politicians or officials (scientists and environmentalists corresponded to 5–8% each). Costa's (2006) research in the main Brazilian newspapers and news magazines between 1975 and 2002 (mainly focused on deforestation and forest burning) confirmed the prime position of governmental sources throughout three decades, followed by research-related sources and NGOs, especially foreign/international ones in both cases. In contrast, says Costa, 'smallhold farmers, landowners, indians, loggers and their respective representation bodies appeared in secondary positions (…) either as villains or as victims of the marked deforestation and forest burning processes in the Amazon' (p. 53) (our translation).

In ANDI's study, national, foreign, and international governmental sources taken together, amount to 35.6% of sources cited in 2005–2007 and 37.2% in 2007–2008 (Vivarta, 2010, p. 43). Based on several indicators, they concluded that 'coverage tends to be presented to readers from the government's outlook' (p. 53) (our translation). Rodas and Di Giulio's recent analysis (2017) of coverage of climate change and energy issues in *Folha de São Paulo* for a decade and a half (2000–2014) confirmed the prominence of governmental sources, which they associated with a news focus on political decisions and on disputes staged in international meetings on climate change.

Contrasting with the press, one of the very few studies of television representations of climate change, which looked at Globo's *Fantástico* show (an infotainment programme), found that scientists were the most widely present sources, followed by members of the public and government representatives (Scalfi, Massarani, Ramalho, & Amorim, 2013).

Following the Scientific Consensus

Brazilian media's focus on the scientific consensus on climate change, as represented by the IPCC, has been pointed out in various studies. ANDI's extensive study showed that the consensus view was dominant and that the vast majority of news stories referred to the link between climate change and human activity: 59.4% in 2005–2007 and 63.9% in 2007–2008 (Vivarta, 2010, p. 41). Painter and Ashe (2012) found that climate scepticism was almost non-existent in the Brazilian press, which featured the lowest percentage of six countries (Brazil, China, France, India, UK, and US).

In a more recent study, Dayrell and Urry (2015) examined a corpus of nearly 20,000 news texts published in 12 daily newspapers between 2003 and 2013 with corpus linguistic techniques. They reported finding 'little evidence of climate change scepticism over the previous decade. (…) Overall, the mainstream Brazilian media adopted, organized and mobilized a gradualist discourse' (p. 269). The 'gradualist' view would be the one represented by the IPCC reports—that as climates are changing due to human activities, economies can be adjusted to reduce future temperature increases (p. 259)—and is contrasted with discourses of 'scepticism' and 'catastrophism'. Exploratory research suggests that some television programmes may offer a more alarmist picture (Matos, Santos, Alvarenga, & Sales, 2012).

(Re)producing Top-down Techno-managerial Discourses

In analysing the coverage of COP15 and COP16 in Brazil's largest news magazines—*Veja*, *Isto É*, Época and *Carta Capital*—Girardi et al. (2013) found a significant emphasis on economics, namely the costs involved in addressing climate change, which was constructed discursively as a matter of negotiation, like 'a business desk of governments and international businessmen who judge how much they can "lose" if a rapid reduction of greenhouse gases is needed' (p. 191). The researchers noted that 'the kind of journalism observed in [their] analysis [could] not break with this [ecotechnocratic] logic' (Girardi et al., 2013) (our translation). Arguably, in this type of representation climate change becomes a political and economic dispute and is further retrieved from citizens' realm.

Reis (1999, p. 150) had already noted a strong presence of economic topics in stories on the 1992 summit, which he associated with the 'Brazilian's media then prevailing opinion that environmental problems cannot be solved, or even mitigated, without massive investment (which, in the case of developing nations, should come mainly from industrial countries)'. In the run up to and during Rio+20, the 2012 UNCED summit, the theme of the 'green economy' gained hegemonic status in the Brazilian media and worldwide (Girardi et al., 2012; Moraes, 2015), further normalizing techno-managerial views. The concept of 'green economy' comes from the sustainable development discourse with a promise of continuous economic improvement derived from environment-oriented policies and technologies, and strategic ambiguity regarding environmental impact and natural resource availability (Kenis & Lievens, 2015).

Moraes (2015; 2017) examined discourses on climate change in Brazilian news magazines during Rio+20. While she did not encounter totally monolithic constructions and a few threads of alternative thinking were weaved in representations of climate change, the discursive edifice rested solidly on economic priorities. Other media portrayed climate politics along similar lines. Zamith et al. (2015, p. 351) found that *Folha de São Paulo* was 'more likely to promote frames that failed to challenge the status quo' than newspapers from countries with smaller contributions to climate change than Brazil's. They went on to argue that its readers …

> … were arguably presented with a normative construction that simultaneously fails to suggest choices for future direction or change and instils a sense that individual or collective action leading to real change may not be realistically feasible. Therefore, readers of the *Folha* (…) may come to see the issue as a nuisance that is both complex and prohibitively expensive to address and thus demand less action from their government. (p. 351)

Looking Away from Meat—and Other Silences

Most analyses of media coverage tend to focus on what is present and overlook what is not. However, silences can have crucial implications, including the indirect legitimation of given practices. Despite the potential effectiveness of a serious reduction of meat consumption (especially beef), and multiple co-benefits in terms of health, environment and food security, that idea continues to be a near taboo for most journalists and other social actors. Noting that at least half of Brazilian emissions result from meat production (and up to 80% from the related categories of land use change and agriculture), Lahsen (2017) has recently produced a striking analysis of the press avoidance of this issue. She analysed the distribution of references to meat and alternative energies in climate-mentioning articles in *Folha de São Paulo* and *O Estado de São Paulo* and found an 'overwhelming focus on energy issues, unreflective of national emissions profile' (p. 23). In sharp contrast, the contributions of meat to Brazilian greenhouse gas emissions were hardly ever present in the press and when mentioned they tended to be downplayed. The role of meat in Brazilian (food) culture and the power of the agribusiness sector (including at the political level) are among the factors that may explain this journalistic blind spot.

The ANDI study (Vivarta, 2010) pointed to other aspects that could be improved or more frequently addressed. They found that climate change coverage is 'highly concentrated' (p. 7) in the newspapers with largest circulations and rarely features in the agenda of smaller local newspapers, which could be important to draw bridges with the spaces and realities that people inhabit. Whereas they found a move in the press coverage from climate impacts to responses, there was a low level of critical analysis of the causes of climate change and of political decisions. They argue that discussion should be expanded 'contemplating not only technical aspects related to [climate change], but also placing issues related to politics, economics and behaviour' at the core of the debate (p. 62). Relatedly, Amaral and Rubin (2013) pointed to the common disconnect between issues in the press. Links between climate change and other problems, such as urban planning or river management and associated political responsibilities, although crucial were not discussed in the issues of *Zero Hora* newspaper that they analysed.

News-making and News Consumption

Several environmental communication scholars have called for research integrating the different moments of the news circuit (Hansen, 2011; Olausson & Berglez, 2014). Studies that simultaneously consider news-making contexts, conditions, and practices (ideally including analysis of the discourses of relevant social actors); media(ted) discourses; and the processes of consumption of news about a given issue are likely to offer a clearer picture of the processes of encoding and decoding of meaning and of how news functions socially.

Together with their analysis of news texts, Rodas and Di Giulio (2017) interviewed journalists from several Brazilian media, who judged the national coverage of climate change as generally 'superficial'. The emphasis on high-level international meetings was justified with a stated preference to report on 'important people's' statements (p. 112) (our translation). Other judgements about the public's values and expectations appear to impact on journalistic decisions. For instance, one journalist noted that while 'people claim that the media should have a preventive approach' the 'Brazilian culture is not like that (...) Brazilians want to know about a crisis when it is already taking place' (p. 112) (our translation). Journalists acknowledged that they were rather reliant on international news agencies, which was partially explained with a lack of human resources. The study also suggested that timeliness (with coverage often tied to key events), human interest, and conflict were the dominant news values that journalists considered in covering climate change.

Loose (2016) combined her analysis of a local newspaper's coverage with a study of perceptions of journalists, news sources and readers regarding climate change and attempted to understand the socio-cultural circulation of climate change discourses. In interviews with journalists of *Gazeta do Povo*, a newspaper from Curitiba, Paraná state, about their work on climate change, she identified an important factor external to newsmakers, namely the availability of information sources. In interviews, journalists held that local sources were 'less accessible' than national-level officials and other 'central' sources, who answer press requests regularly. While observing a newsroom's routines, she found that journalists would think primarily of national reference figures when producing news stories on climate change and discarded the possibility of contacting experts from the newspaper's hometown due to difficulties experienced in previous attempts. When Loose interviewed potential local sources they admitted that they did not have a pro-active attitude towards the media. She concluded that local policy fora and locally-based scientists did not offer ground material for journalists to build the global-local relation involved in climate change. Another important point refers to journalists' conceptions about climate change: structural aspects such as consumerism and the capitalist system were less present in their discourse than research suggests is the case with journalists from other countries (Loose & Carvalho, 2015; cf. Engesser & Brüggemann, 2016).

Turning to news readers, Loose (2016) found that views on climate change were disconnected from daily life in Curitiba as the media's focus on negative global effects and international treaties had repercussions in people's understanding of the problem. Inquired readers pointed out gaps in the coverage of climate change in the local newspaper *Gazeta do Povo*, especially in terms of local significance.

Conclusions and Future Research

Brazil has the potential to play a key role in the politics of climate change given its natural resources, sheer size, and several other factors. Whereas the country rose to prominence in the international stage for good reasons from around 2009, the last few years have seen a series of setbacks in climate policies and other related measures. The decline in political commitment stands in contrast with a positive 'climate of opinion' as public support for action is very high. Brazil faces large risks from climate change impacts and is highly vulnerable, which calls for more and better resilience-oriented policies. Appropriate policies and implementation mechanisms for forestry, agriculture, livestock, and other land uses, among other sectors, could also contribute strongly to reduce greenhouse gas emissions and mitigate climate change.

In a context of limited journalistic freedom and resources, climate change has mainly been put on the agenda via international events and international news agencies, and mostly depicted as global and distant issue, although some research suggests that the focus on national dimensions of climate change has been growing in recent times. Official sources have been clearly dominant in the press and media discourse has been predominantly techno-managerial, perhaps unsurprisingly in pro-neoliberalism media. The strong ties between media organizations and economic and political interests in Brazil are likely to also play a role in the exclusion of more transformational discourses. Nonetheless, when various factors are taken into account, the Brazilian press emerges as favourable to government-led mitigation policy, which compares positively with other countries' media (Broadbent et al., 2016).

The first obvious observation about research gaps and opportunities on media and climate change in Brazil concerns the relative lack of analyses of television. Although this is not unique of Brazil, it is all the more significant given the place of television in the national culture. Analyses of televised news as well as of representations of climate-relevant matters in other television genres would certainly help understand the symbolic construction of climate change in the country. Radio also deserves more attention from researchers.

Another gap concerns alternative (or independent) news media. Although with small audiences, those are the platforms where status quo-defying discourses are most likely to be found. If indeed those discourses are present, how they are developed, how they are received and how they play into the engagement of certain social groups would be valuable research questions (Carvalho, van Wessel, & Maeseele, 2017).

Future research would also benefit from expanding in terms of scope. Most of the studies on Brazil (as elsewhere) have consisted of content (or discourse) analyses of media coverage. Although it is important to continue examining the evolution of media representations of climate change, it is vital to know how different publics consume/interpret/appropriate those representations. Studying the full cultural circuit of climate change would also involve examining how different social actors communicate about it and how journalists filter and reconstruct those discourses.

References

Albuquerque, A. (2012). On models and margins: Comparative media models viewed from a Brazilian perspective. In D. C. Hallin & P. Mancini (Eds.), *Comparing media models beyond the western world* (pp. 72–95). Cambridge: Cambridge University Press.

Amaral, M. F., & Rubin, A. (2013). O valor das notícias no jornal brasileiro Zero Hora. *Razón y Palabra, 18*(84). Retrieved from http://www.redalyc.org/articulo.oa?id=199528904010

Angelo, C. (2016). O primeiro desastre ambiental da era Temer? *Observatório do Clima.* Retrieved from http://blog.observatoriodoclima.eco.br/?p=2695#sthash.aRy7Cz8F.dpbs

Börner, J., Wunder, S., Wertz-Kanounnikoff, S., Tito, M. R., Pereira, L., & Nascimento, N. (2010). Direct conservation payments in the Brazilian Amazon: Scope and equity implications. *Ecological Economics, 69*(6), 1272–1282. doi:10.1016/j.ecolecon.2009.11.003.

Broadbent, J., Sonnett, J., Botetzagias, I., Carson, M., Carvalho, A., Chien, Y.-J., … & Zhengyi, S. (2016 January–December). Conflicting climate change frames in a global field of media discourse. *Socius: Sociological Research for a Dynamic World, 2.* doi:10.1177/2378023116670660

Carvalho, A., van Wessel, M., & Maeseele, P. (2017). Communication practices and political engagement with climate change: A research agenda. *Environmental Communication, 11*(1), 122–135. doi: 10.1080/17524032.2016.1241815

Costa, L. M. (2006). O esverdeamento da imprensa. *Estudos em Jornalismo e Mídia, 3*(2), 41–54. Retrieved from https://periodicos.ufsc.br/index.php/jornalismo/article/download/2289/2017.

Dayrell, C., & Urry, J. (2015). Mediating climate politics: The surprising case of Brazil. *European Journal of Social Theory, 18*(3), 257–273. doi:10.1177/1368431015579962

Dunlap, R. E. (1998). Lay perceptions of global risk. Public views of global warming in cross-national context. *International Sociology, 13*(4), 473–498.

Engesser, S., & Brüggemann, M. (2016). Mapping the minds of the mediators: The cognitive frames of climate journalists from five countries. *Public Understanding of Science, 25*(7), 825–841. doi:10.1177/0963662515583621

Fioravanti, C. (2007). *Climate change reporting in Brazil. Presentation at workshop Carbonundrums: Making sense of climate change reporting around the world.* Oxford, UK. University of Oxford. Retrieved from http://www.eci.ox.ac.uk/news/events/070727-carbonundrum/fioravanti.pdf

Fonseca, F. (2010). *Mídia e poder: Elementos conceituais e empíricos para o desenvolvimento da democracia brasileira.* Texto para discussão 1509. Brasília: Instituto de Pesquisa Econômica Aplicada. Retrieved from http://www.ipea.gov.br/portal/index.php?option=com_content&view=article&id=9660

Friedrich, J., De, M., & Damassa, T. (2015). Infographic: What do your country's emissions look like? Retrieved from http://www.wri.org/blog/2015/06/infographic-what-do-your-countrys-emissions-look

Girardi, I. M. T., Massierer, C., Moraes, C., Loose, E. B., Neuls, G., ... & Gertz, L. (2013). Discursos e vozes na cobertura jornalística das COP15 e 16. *Em Questão, 19*(2), 176–194. Retrieved from http://seer.ufrgs.br/index.php/EmQuestao/article/view/28599

Girardi, I. M. T., Moraes, C., & Loose, E. B. (2012). Bases do jornalismo ambiental e os desafios para a cobertura da Rio+20. *Razón y Palabra, 79*, 1–29. Retrieved from http://www.razonypalabra.org.mx/N/N79/M79/01_TourinhoHerteBeling_M79.pdf

Hansen, A. (2011). Communication, media and environment: Towards reconnecting research on the production, content and social implications of environmental communication. *International Communication Gazette, 73*(7), 7–25. doi:10.1177/1748048510386739

Howard, B. C. (2014). Brazil leads 2014-world in reducing carbon emissions by slashing deforestation. *National Geographic.* Retrieved from http://news.nationalgeographic.com/news/2014/06/140605-brazil-deforestation-carbon-emissions-environment/

Instituto Nacional de Pesquisas Espaciais. (2016). PRODES estima 7.989 km^2 de desmatamento por corte raso na Amazônia em 2016. Retrieved from http://www.inpe.br/noticias/noticia.php?Cod_Noticia=4344

IPCC (Intergovernmental Panel on Climate Change). (2014). *Climate change 2014: Impacts, adaptation and vulnerability.* Retrieved from https://www.ipcc.ch/report/ar5/wg2/

Kenis, A., & Lievens, M. (2015). *The limits of the green economy.* London: Routledge.

Lahsen, M. (2017). Buffers against inconvenient knowledge: Brazilian newspaper representations of the climate-meat link. *Desenvolvimento e Meio Ambiente, 40*, 17–35. doi: 10.5380/dma.v40i0.49002

Loose, E. B. (2016). *Riscos climáticos no circuito da notícia local: Percepção, comunicação e governança* (Doctoral dissertation). Retrieved from http://acervodigital.ufpr.br/handle/1884/43179.

Loose, E. B., & Carvalho, A. (2015). O que pensam os jornalistas sobre seu papel para o enfrentamento das mudanças climáticas: As percepções dos profissionais da Gazeta do Povo no Paraná, Brasil. *Razón y Palabra, 91*, 36 online, n.p. Retrieved from http://www.razonypalabra.org.mx/N/N91/Monotematico/02_BelingCarvalho_M91.pdf

Matos, H., Santos, E., Alvarenga, J., & Sales, M. C. L. (2012). O discurso da TV sobre o aquecimento global a partir da série de reportagens 'Terra, que tempo é esse?' do jornalístico Fantástico. *Revista Geonorte. Edição especial, 2, 1*(5), 642–654. Retrieved from http://www.periodicos.ufam.edu.br/revista-geonorte/article/view/2401/2222

Ministério do Meio Ambiente. (2012). O que o brasileiro pensa do meio ambiente e do consumo sustentável—Edição 2012. Retrieved from http://hotsite.mma.gov.br/redemulheres/wpcontent/uploads/Sumario-Executivo_MMA_port_25jun2012.pdf

Moraes, C. H. (2015). *Entre o clima e a economia: enquadramentos discursivos sobre a Rio+20 nas revistas Veja, Isto É, Época e Carta Capital* (Doctoral dissertation). Retrieved from https://www.lume.ufrgs.br/bitstream/handle/10183/114650/000955664.pdf?sequence=1

Moraes, C. H. (2017). A mudança climática no enquadramento discursivo da revista Época. *Desenvolvimento e Meio Ambiente*, 40: 37–46. doi:10.5380/dma.v40i0.49071.

Observatório do Clima. (2016). Emissão do Brasil sobe 3,5% em 2015, mostram dados do SEEG. Retrieved from http://www.observatoriodoclima.eco.br/emissoes-do-brasil-sobem-35-em-2015-mostra-seeg/

Olausson, U., & Berglez, P. (2014). Media and climate change: Four long-standing research challenges revisited. *Environmental Communication, 8*(2), 249–265. doi:10.1080/17524032.2014.906483

Oliveira, D. (2011). Jornalismo alternativo: Um potencial para a radicalização da democracia. *Signo y Pensamiento, 30*(58), 52–63. Retrieved from http://www.scielo.org.co/scielo.php?pid=S0120-48232011000100004&script=sci_abstract

Painter, J., & Ashe, T. (2012). A cross-national comparison of the presence of climate scepticism in the print media in six countries, 2007–10. *Environmental Research Letters, 7*(4). doi:10.1088/1748-9326/7/4/044005

PBMC—Painel Brasileiro de Mudanças Climáticas. (2013). PBMC. Contribuição do Grupo de Trabalho 2 ao Primeiro Relatório de Avaliação Nacional do Painel Brasileiro de Mudanças Climáticas. Sumário Executivo do GT2. Retrieved from http://www.pbmc.coppe.ufrj.br/documentos/MCTI_PBMC_sumario_executivo_impactos_vulnerabilidades_e_adaptacao_WEB_3.pdf

Pew Research Center. (2007). *Global unease with major world powers.* Retrieved from http://www.pewglobal.org/2007/06/27/global-unease-with-major-world-powers/

Pew Research Center. (2015). *Global concern about climate change, broad support for limiting emissions.* Retrieved from http://www.pewglobal.org/files/2015/11/Pew-Research-Center-Climate-Change-Report-FINAL-November-5-2015.pdf

Reis, R. (1999). Environmental news: Coverage of the Earth Summit by Brazilian newspapers. *Science Communication, 21*(2), 137–155.

Rodas, C., & Di Giulio, G. (2017). Mídia brasileira e mudanças climáticas: uma análise sobre tendências da cobertura jornalística, abordagens e critérios de noticiabilidade. *Desenvolvimento e Meio Ambiente, 40*, 101–124. doi: 10.5380/dma.v40i0.49002

Scalfi, G., Massarani, L., Ramalho, M., & Amorim, L. (2013). Mudanças climáticas em um programa infotainment: Uma análise do Fantástico. *Razón y Palabra, 18*(84), Retrieved from http://www.redalyc.org/articulo.oa?id=199528904006

Secretaria da Comunicação Social da Presidência da República. (2016). Pesquisa brasileira de mídia 2016. Retrieved from http://www.secom.gov.br/atuacao/pesquisa/lista-de-pesquisas-quantitativas-e-qualitativas-de-contratos-atuais/pesquisa-brasileira-de-midia-pbm-2016.pdf.

Vergara, W., Rios, A. R., Galindo Paliza, L. M., Gutman, P., Isbell, P., Suding, P. H., & Samaniego, J. (2013). *The climate and development challenge for Latin America and the Caribbean: Options for climate-resilient low-carbon development*. Washington, DC: Inter-American Development Bank. Retrieved from https://publications.iadb.org/handle/11319/456

Viola, E. (2004). Brazil in the context of global governance politics and climate change, 1989–2003. *Ambiente & Sociedade, 7*(1), online, n.p. doi:10.1590/S1414-753X2004000100003

Viola, E. (2013). Transformations in Brazilian deforestation and climate policy since 2005. *Theoretical Inquiries in Law, 14*, 109–123. Retrieved from http://www7.tau.ac.il/ojs/index.php/til/article/view/5

Vivarta, V. (Ed.). (2010). Mudanças climáticas na imprensa brasileira: uma análise comparativa de 50 jornais nos períodos de julho de 2005 a junho de 2007- julho de 2007 a dezembro de 2008 Brasília, ANDI. Retrieved from http://www.andi.org.br/sites/default/files/Mudanca%20climatica%20na%20imprensa%20brasilerira%202005%202008.pdf

Zamith, R., Pinto, J., & Villar, E. (2015). Constructing climate change in the Americas: An analysis of news coverage in U.S. and South American newspapers. *Science Communication, 35*, 334–357. doi:10.1177/1075547012457470

6. *Environmental Journalism in China: Challenges and Prospects*

Sibo Chen

Introduction

Questions concerning the past, present, and future of natural environment in China are important for us all. Domestically, the country's rapid economic development and industrialization since the 1980s has dramatically increased its consumption of natural resources and generated horrific ecological impacts, which in turn give rise to escalating environmental crises nationwide (Shapiro, 2012). Internationally, the country has received both accolades and condemnations from other countries for its ambivalent role in international environmental politics. As the world's second largest economy and top emitter of greenhouse gases (GHG), China has become a crucial, or even decisive, player in multilateral climate change negotiations. On the one hand, China's firm commitment to renewable energy innovation and its executive implementation of environmental policies have impressed many Western commentators. It is the first among the major industrial countries to expressly commit to pursuing a vision of 'ecological civilization' and this vision has been confirmed in the Communist Party of China's (henceforth 'the Party') constitution since 2012 (Wang, He, & Fan, 2014). On the other hand, as the world's top GHG emitter, China is reluctant to take more responsibility in reducing its GHG emissions and this position has been strongly criticized by other industrial countries (Zhang & Barr, 2013). For instance, the previous conservative government of Canada repetitively used China as an excuse for its own inaction on climate change.

The media system is a contested domain for discursive struggles over China's ecological future. Cox (2013) argues that the production of environmental stories is inevitably influenced by media political economy, media routines, and other factors such as journalist norms and news frames. Through various media outlets, different news stories and commentaries shape and influence ordinary citizens' perceptions of and attitudes toward the country's enduring environmental crises. Yet, the evolving environmental journalism in China, despite its indispensable role in domestic and international environmental politics, has received scarce attention within the research literature. Although in recent years this situation has begun to change, with research being conducted on topics such as journalistic authority and environmental journalists (e.g., De Burgh & Rong, 2011 Dong, 2013; Hung, 2013; Tong, 2014, 2016), little attention has been paid to how meanings of climate reporting are negotiated in discursive spaces constructed by China's media political economy.

This chapter sets out to fill the above gap through a case study of two Chinese media's coverage on the 2015 United Nations Climate Change Conference (COP 21). The overall aim is to elucidate how climate reporting is characterized by the complexity of socio-political factors in China. It argues that domestic media tend to frame climate change from economic and geopolitical perspectives, thereby legitimizing the Chinese government's aspiration for global environmental leadership. To further delineate this argument, let us begin with an overview of China's complex ecological situation.

Previous Research and Theory

Although some scholars argue that China's environmental degradation can be traced back to Mao's era or even the late Qing dynasty, the current environmental crises it suffers are largely rooted in its extraordinary pace of economic development and industrialization since the late 1980s (Shapiro, 2012). A process of modernization that took over two centuries in Europe and North America has been accomplished by China in less than 30 years. Consequently, the extent, magnitude, and intensity of its environmental crises have been much more dramatic than those in North America and Europe, leaving little time for the Chinese government and people to respond. Besides being the world's top GHG emitter, currently China is home to 16 of the world's 20 most polluted cities (World Bank, 2014). Although the country has one of the largest fresh water resources in the world, it is facing serious water shortages due to pollution, inefficient usage, and the skyrocketing demands of industry and agriculture. Its biological diversity and reserve of arable land are also under constant threat.

According to Shapiro (2012), current environmental crises in China are primarily driven by five factors: population increase, rapid industrialization, the burgeoning middle class, urbanization, and globalization. While these factors also perplex other developing countries, their magnitude and complexity are often exaggerated by China's unique economic, social, and political conditions. Take the country's burgeoning middle class as an example, while its demand for a cleaner, healthier environment does create public pressure upon the government, in many cases such pressure only leads to the relocation of toxic industries to poor rural areas, leaving vulnerable communities to suffer the consequences of environmental inequality. Another telling example is the country's rapid urbanization process. The government's policy preference toward urban areas creates the public perception that rural residents are inferior to their urban peers. 'Being urbanized' becomes an enhancement of social status. As a result, contemporary China witnesses a huge rural-urban gap that puts environmental pressure on both sides: while booming metropolitans are caught up in increasing density and traffic, declining villages become victims of agricultural contamination and illegal dumping of industrial waste.

Taken together, contemporary China confronts a catastrophic collapse of its ecological system. The situation is so severe that it not only threatens the country's public health and future prosperity, but also presents an acute political challenge to China's central and local governments (Shapiro, 2012; Wang et al., 2014). In this context, the purposes of environmental journalism are twofold: while some reports provide crucial legitimacy for the government's efforts on environmental protection, others challenge the hegemonic logic of economic modernization that remains strong in policy practices. Domestic environmental reporting thus tends to focus on issues with visible and direct threats to public health, which makes it difficult to fit the pervasive yet largely invisible environmental risks of climate change into the Chinese media's regular news cycle. In this sense, climate reporting's relative periphery in China's media system is a direct outcome of the complexity of the country's ecological challenges. This situation only changes during climate summits, such as the 2015 United Nations Climate Change Conference (COP 21).

Meanwhile, environmental journalism is also characterized by economic and political concerns underlying China's hegemonic social structures. Tong (2014, 2016) argues that environmental journalism in China is constrained by (1) the government's prioritization of economic growth, (2) the decentralization of environmental governance, (3) weak public participation, (4) political control over civil organization, and (5) the public's weak perception of and behaviour toward environmental issues. Although Tong's interpretation still follows the liberal approach that considers China's media dynamic in terms of 'control versus resistance', she correctly points out the political nature of environmental reporting in the Chinese context. Historically, issues such as industrial pollution and over-exploitation of natural resources were simply neglected or regarded as the 'inevitable price of prosperity' (Zhang & Barr, 2013). More recently, certain environmental stories were blocked by media censorship because they were deemed as either humiliations of national face or threats to social stability.

Despite these constraints, however, China has witnessed a steady growth of domestic environmental reporting over the past decade, which is primarily driven by two factors. First, the increasing visibility of environmental issues in daily life makes it difficult to maintain the outdated 'pollution first, treatment later' narrative in environmental governance. Second, the Chinese government (especially its central administration) allows the exposure of environmental problems to some extent, especially when it considers such exposure as politically safe or as necessity for the Party to communicate to the public about its determination and ability to tackle environmental challenges.

Given the international nature of climate negotiations, Chinese media coverage of them is also driven by the country's rising global status. As Zhao (2014) argues, the evolving global power shifts, as marked by a transition from unipolar American hegemony to a multi-polar global order, have made the racial logic of global capitalism intersect with the nation-state and class dimensions of power. In the realm of global communication, the complexity of such intersections often prevails in nation-state-centric narratives fuelled by nationalistic expressions. In the Chinese context, the discursive domain occupied by media outlets is the meeting ground of state-sponsored nationalism and populist reactive nationalism (Lee, 2003; Sandby-Thomas, 2011). On the one hand, the Party's political rhetoric has gradually turned to economic development and patriotic nationalism to legitimize its governance within the neoliberal world order. As Chan (2001) notes, when China receives global recognition for its rising international status or its sovereignty is infringed by foreign forces, propaganda-weary netizens and even some overseas dissidents rally behind the Party's official narratives. In respond to such tendencies, the Party's political rhetoric acknowledges and encourages the celebration of the 'China Dream' among the public. On the other hand, the rise of patriotic nationalism should also be read as the Chinese public's reaction against neoliberal globalization. The incorporation of transnational capital into the country's different economic sectors, has primarily served the interests of the elite groups, which 'aggravates the political economic and cultural contradictions of China's re-articulation within an American-dominated global capitalism' (Zhao, 2008, p. 166).

In this context, patriotic nationalism becomes a crucial and sometimes even decisive factor influencing domestic media narratives on the Chinese government's standpoints in international affairs, including multilateral climate negotiations. These narratives are often built around whether government decisions enhance national face or protect national sovereignty, rather than whether they are leaning toward global ecological sustainability. One telling example of this tendency is how domestic media rally behind the Chinese government's ambiguous standpoint on GHG emissions. The Chinese government insists that GHG emissions should be judged on a per capita basis and as a developing country, and that China's commitment to fighting against climate change should not hinder its economic development. Such a standpoint has been contested by many other industrial countries. When the country is criticized as a 'climate villain' on the international stage, however, the strong reactions from domestic media ignite sentiment among patriotic netizens, which in turn breeds a unique form of 'climate scepticism'. This scepticism focuses less on the validity of climate science; instead, it emphases the unfairness of international climate deals, presenting them as new measurements from 'foreign enemies' to disrupt China's thriving global influence. Some popular texts even adopt the term 'carbon colonialism' to cater to the public's feeling that the critiques on the country's massive GHG emissions are direct insults on its national face. In short, the interplay between patriotic nationalism and media discourse plays a key role in framing how the Chinese government's approach to climate negotiations is communicated to domestic audience. Indeed, when Xie Zhenhua (China's special representative to climate change talks) threw back the European Union's criticism on China with the words 'what entitles you to lecture me here' during the Copenhagen Summit back in 2009, he expressed the outrage and frustration shared by many ordinary citizens (Zhang & Barr, 2013). Confronting the unfairness in many global carbon-reduction proposal and the hypocrisy of developed countries, such sentiment reflects China's frustration with its status in global politics and the nation's bitter memory of its diplomatic failures before the establishment of the People's Republic of China.

Overall, for Chinese media, climate change presents a subject with embedded ambiguities, contradictions, and tensions. The challenges and opportunities created by such complex profoundly influence the ways in which climate change is communicated domestically. My argument here is not meant to advocate a sense of 'Chinese exceptionalism'. Instead, my point is that only through examining the country's escalating environmental degradation in terms of its gradual integration into the global capitalist system can we reach a comprehensive recognition of the relationship between its environmental reporting and ecological status quo.

Reading Chinese Media's Representations of COP 21

It is useful to explore Chinese media's coverage on COP 21, arguably the milestone of international climate negotiations since the Kyoto Protocol. While this analysis cannot do justice to the many facets of Chinese environmental journalism, such an overview still offers insights into how media's domestication of climate change is intertwined with economic and geopolitical concerns in the Chinese context.

The current case study is a qualitative analysis of a sample of 41 articles published during a 30-day period (November 21 to December 20, 2015) from two media outlets: *People's Daily* and *China Daily*. As the official newspaper of the Communist Party of China, *People's Daily* sets the public agendas of domestic and international politics. A distinctive feature of this paper is its commentary section, which is often collectively authored and represents the government's official standpoint. By comparison, *China Daily* is owned by the Chinese government's state council information office. As the most-read English language newspaper published in China, it focuses more on communicating a 'rising China' to both domestic and an international English-language audience. Although *China Daily* claims no connection with *People's Daily*, the two papers often cross-publish each other's articles and their ideological stances as well as editorial styles are identical. The selected articles include news reports and commentaries on COP 21 as well as other reports on international relations, energy issues, and environment that explicitly mention or discuss the Chinese government's climate change policies.

The overachieving analytical framework is Fairclough's (2010) critical discourse analysis. Fairclough considers oral and written discourses as social actions intertwining with the real world. Accordingly, the primary goal of this analysis is to elucidate how climate change is discursively constructed in relation to the Chinese government's broad economic and political concerns. For this reason, the selected articles are treated as a collective: the chronological account below will look at how climate change is addressed before, during, and after COP 21.

Table 6.1 shows the distribution of the selected articles between the 30-day period. The overall pattern presents two notable trends. First, among the 41 articles analyzed, *China Daily* published 33 of them. Second, while *China Daily* was very responsive to the global attention caused by COP 21, devoting 15 reports on it between December 1 and December 10, climate reporting remained steady in *People's Daily* during the 30-day period. Such discrepancies do not imply that *China Daily* had considerably more attention to climate change. In contrast to *People's Daily*, *China Daily* is more willing

Table 6.1: Coverage on climate change and COP by *People's Daily* and *China Daily*.

	People's Daily	*China Daily*
November 21–30	3	9
December 1–10	2	15
December 11–20	3	9

to accommodate third-party voices due to its primary audience (English-language readers inside and outside China). In fact, a considerable proportion of the articles from *China Daily* are invited columns written by guest authors (intellectuals, government officials, etc.). Yet, unlike the balanced reporting approach followed by agenda-setting media in some other countries, these 'guest voices' form a unified storyline complementing the official narrative set by both media outlets: as a rising global power, China is playing an active and responsible role in climate politics.

To begin with, a notable argument made by the articles from *China Daily* during the days leading to COP 21 was that climate change is a clear, near-term threat to the world, especially developing countries. What repeatedly appears in these articles were position statements urging developed countries to take more responsibility by honouring their commitments to offer financing and technology to developing countries. In one report covering the Chinese President Xi Jinping's planned deliberation at COP 21, China's expectation was described thus:

> Liu (Liu Zhenmin, Vice-Foreign Minister China) said China expected the climate summit to yield a 'comprehensive and balanced' agreement. The conference should push for more concrete actions from developed countries, he said, adding that these countries are expected to honour their commitments to offer financing and technology to developing countries. 'The principles of equity, common but differentiated responsibilities and respective capabilities should be highlighted, since they reflect the fact that developed countries have made a greater historical contribution to climate change,' Liu said.
>
> 'Xi to deliver climate change plan in Paris'
> *China Daily*, November 26, 2015

Statements like this presented a critical stance against the inaction of developed countries, often juxtaposed with a factual paragraph or sentence explaining the Chinese government's achievements in curbing its GHG emissions. This binary construction highlighted the political tension between developing and developed countries. Accordingly, carbon pricing and green technology were recommended as the key solutions to rescue the world out of the current stagnation in climate politics.

> Equally important is global businesses' support for an ambitious deal in Paris that could give a huge boost to their confidence to invest heavily in long-term green growth. Also on Monday, the chief executives of 78 major companies, with combined annual turnover of $2.1 trillion, issued an open letter urging world leaders to seize the chance to strive for an ambitious deal featuring effective climate policies that would trigger low-carbon investment and transform current emission patterns at a significant scale to help create both economic growth and jobs.
>
> 'An ambitious deal in Paris is needed to boost green growth'
> *China Daily*, November 26, 2015

To strengthen its urge for climate justice, *China Daily* also published opinion pieces by leaders of world organizations. In a commentary authored by Jim Yong Kim (President of the World Bank Group) and Christine Lagarde (Managing Director of the International Monetary Fund), the two leaders expressed their support for global carbon pricing, a proposal supported by the Chinese government.

> The transition to a cleaner future will require both government action and the right incentives for the private sector. At the center should be a strong public policy that puts a price on carbon pollution. Placing a higher price on carbon-based fuels, electricity and industrial activities will create incentives for the use of cleaner fuels, save energy and promote a shift to greener investments. Measures such as carbon taxes and fees, emissions-trading programs and other pricing mechanisms, as well as removal of inefficient subsidies can give businesses and households the certainty and predictability they need to make long-term investments in climate-smart development.
>
> 'Path to carbon pricing for a green world'
> *China Daily*, November 24, 2015

Compared with the advocacy role taken by *China Daily*, *People's Daily* was more prudent in explicating China's position prior to COP 21. Between November 21 and 30, it only published one article addressing the government's hope for a comprehensive and balanced agreement at COP 21. The most interesting aspect of this carefully written position statement, however, was how it took climate change as a confirmed scientific reality by linking China's domestic environmental challenges to the global efforts on GHG emission reduction. This link was also made in the other two pieces published by *People's Daily* during this period, with one discussing the abnormal winter weather coming to China and the other addressing the fact that the country's temperature and sea levels were rising faster than global average. The affirmative tone of these reports indicated that China's official discourse on climate change seems to move beyond the issue of 'scientific consensus'. The discussions of alarming weather patterns also resonated with China's frustration over the challenge of handling development and environment protection simultaneously. As *China Daily* wrote:

> Over the years, heavy smog in many parts of China has been ringing alarm bells to the nation's mode of economic and social development. The Chinese have realized the importance of protecting the environment as well as maintaining sustainable development, in light of which, cutting carbon emissions is closely related to reducing smog. The Chinese society is more anxious about the carbon emission reduction than the outside world, which has been busy criticizing China for generating too much emissions. Yet it won't be easy to eliminate smog and promptly cut down the carbon emissions, given the difficult balance that China has been trying to keep between development and environment protection over the years.
>
> 'Will Paris climate meet herald a new era'
> *China Daily*, November 30, 2015

The discursive dynamics between *People's Daily* and *China Daily*, with the former offering key position statements and the latter complementing these statements with commentaries and guest pieces, continued during the negotiations at COP 21. Besides expressing the Chinese government's willingness and commitment of making a binding agreement along with other countries, the articles published between December 1 and December 10 also gravitated toward two additional themes. First, China was championing developing countries during the negotiations by pushing its counterpart (especially the US) to act as 'responsible major countries'. Second, what blocked the progress of the climate talks in Paris the inaction of developed countries; China, by contrast, was acting as a constructive player. In line with the media narratives preceding COP 21, both themes emphasized the political nature of climate negotiations and were based on a modest version of climate justice. In one article arguing against the criticism that 'China is blocking the progress of Paris talks', the author wrote:

> China is not blocking the progress of the climate talks in Paris by not being part of a new coalition, but is striving to play a constructive role in the negotiations' crunch time to produce proposals leading to a successful outcome, a Chinese climate expert said. China has positioned itself to play 'a bridging role' between developing and developed countries in the United Nations climate talks, but that doesn't mean it is in a middle place between developing and developed countries (…) The division between developed and developing countries should not be blurred, since the historical responsibilities between developed and developing countries haven't changed (…)
>
> 'China 'not blocking progress' of Paris talks'
> *China Daily*, December 11, 2015

Meanwhile, when discussing fellow developing countries, *China Daily* took a more nuanced approach and discursively posited China at the centre of the world stage. This discursive strategy was frequently adopted in the coverage on Xin Jinping's activities at COP 21:

> President Xi Jinping met with at least 50 world leaders and presented China's solutions to countering global warming and modernizing a less-developed continent, during a weeklong visit to three countries that concluded on Sunday. [...] Wang, a member of Xi's delegation, said the Chinese president had spoken at the Paris climate change conference from a sense of justice for developing countries, pushing for a pact that is fair, reasonable and effective. 'It is imperative to respect differences between countries, especially developing ones, in domestic policies, capacity building and economic structure,' the president said. 'Developing countries' legitimate needs of reducing poverty and improving their peoples' living standards should not be denied while addressing climate change, [...] developed countries should honor their commitment, mobilizing $100 billion each year before 2020, and provide stronger financial support to developing countries afterward.'
>
> 'Xi champions developing nations'
> *China Daily*, December 7, 2015

Claims referring to climate justice by Xi and the Chinese government are interesting in two ways. First, they discursively construct a coalition among the developing countries, in which China acts as the indisputable leader. Second, for domestic audience in China, the leadership implied in such claims fit the overall 'rising China' frame advanced by *China Daily* and *People's Daily*. It is noteworthy that justice claims here are intertwined with the senses of national interest and national identity. Climate change, in this regard, is a form of political power game that China needs to win. This nationally grounded political realism dominates the selected articles.

As the historical Paris Agreement was successfully reached at the end of COP 21, both *China Daily* and *People's Daily* joined other media outlets worldwide to express their cautious optimism. Besides following the themes mentioned above, both media presented a close China-U.S. partnership in climate management as a crucial factor for maintain the good momentum built by COP 21. The depiction of China as *the* counterpart to the US implicitly made a binary construction: the developed world led by the US versus the developing world led by China. Given the extent to which the journalistic field and the political field interact with each other, *China Daily* and *People's Daily* not only share the Chinese government's understanding of climate politics, they are also deeply involved in discursively reinforcing the government's political standpoint.

> As the weeks-long negotiation in Paris entered its final stretch, Chinese President Xi Jinping made a last-ditch effort on Friday with his U.S. counterpart Barack Obama to seek a global deal. In phone talks with Obama, Xi said that China and the United States need to strengthen coordination and work together with all parties to ensure that the Paris climate conference would reach an agreement as scheduled, which would be in the interest of the international community.
>
> 'China urges implementation after historic climate deal sealed'
> *People's Daily*, December 13, 2015

> Global politicians, their teams and advisers in Paris had very good reasons to pack for home in a positive frame of mind, as they have avoided a failure like Copenhagen and successfully reached a legally binding agreement to fight climate woes. They have made this historic moment possible after extensive wrangling over the past two weeks. Chinese president Xi Jinping and US President Barack Obama, as the leaders of the two largest emitters, paved the way for success with their earlier climate commitments. They gave the day-and-night talks a boost with their shared desire to see an agreement reached.
>
> 'Good momentum of climate talks must be maintained'
> *China Daily*, December 15, 2015

Discussion and Conclusion

The articles on COP 21 from *People's Daily* and *China Daily* illustrate how environmental journalism intertwines with China's specific political economy during a moment when climate negotiations are foregrounded as the central focus of international politics. As mentioned earlier, domestic media in China tend to focus on the political nature of environmental issues, using news coverage to legitimize or challenge the Party's environmental governance. In the cases of *People's Daily* and *China Daily*, both media outlets present China as a constructive player and champion of developing countries at COP21, thereby legitimizing the Chinese government's climate policies. This discursive strategy builds upon the senses of stagnation and fragmentation caused by the failures of previous climate summits. Echoing rising nationalism among the public, the strategy also highlights China's aspiration for global environmental leadership, fitting climate change into the broad 'rising China' frame of the Party rhetoric. Meanwhile, the discussions of alarming weather patterns in both media outlets shed light upon the sheer difficulty of keeping a balance between economic development and environmental protection in countries like China, where economic performance is directly connected with political stability.

Although the above findings may not come as a surprise given the extent to which *People's Daily* and *China Daily* follow the Party rhetoric, they still invite us to consider how China's evolving media system and political economy create profound opportunities and challenges for its environmental journalists. The overall pattern found in both media outlets presents a distinctive politicization of climate change (Pepermans & Maeseele, 2016). Unlike many Western media's instances on using scientific evidence to diffuse political polarization and increase consensus, this approach seeks to use climate talks to open new political spaces and challenge the status quo of North-South divide. Admittedly, China's advocacy for climate justice is inevitably influenced by its own ecological crises and unique socio-political pre-conditions. Yet, such call for justice points to questions concerning the embedded unfairness of global economy, which, to many developing countries like China, need to be addressed as we are responding to planetary crises as a collective. The journalistic engagement of Chinese media, in this regard, speaks to the political concerns of many other developing countries.

How shall we understand China's environmental journalism, especially in terms of its implications for the world's environmental future? Clearly, the answer cannot be found in a de-contextualized 'control versus resistance' dichotomy. As shown throughout the chapter, it is more appropriate to conceive the rise of environmental journalism in China as an outcome of the country's complicated socio-economic and political struggles, rather than a simple spread of Western journalist paradigm and environmentalism. China's environmental journalism is, and will continue to be, a site filled by evolving economic, political, social, and cultural struggles.

References

Chan, M. H. (2001). *Return of the dragon: China's wounded nationalism.* Boulder, CO: Westview.

Cox, R. (2013). *Environmental communication and the public sphere.* Thousand Oaks, CA: SAGE Publications.

De Burgh, H., & Rong, Z. (2011). *China's environment and china's environment journalists: A study.* Bristol: Intellect.

Dong, D. (2013). Legitimating journalistic authority under the state's shadow: A case study of the environmental press awards in China. *Chinese Journal of Communication, 6*(4), 397. doi:10.1080/17544750.2013.845110

Fairclough, N. (2010). *Critical discourse analysis: The critical study of language* (2nd ed.). New York, NY: Routledge.

Hung, C. (2013). Citizen journalism and cyberactivism in China's anti-PX plant in Xiamen, 2007–2009. *China: An International Journal, 11*(1), 40–54.

Lee, C. C. (2003). The global and the national of the Chinese media: Discourses, market, technology, and ideology. In C. C. Lee (Ed.), *Chinese media, global contexts* (pp. 1–31). London: Routledge Curzon.

Pepermans, Y., & Maeseele, P. (2016). The politicization of climate change: Problem or solution? Politicization of climate change. *Wiley Interdisciplinary Reviews: Climate Change, 7*(4), 478–485. doi:10.1002/wcc.405

Sandby-Thomas, P. (2011). *Legitimating the Chinese communist party since Tiananmen: A critical analysis of the stability discourse.* Milton Park, Abingdon, Oxon: Routledge.

Shapiro, J. (2012). *China's environmental challenges.* Malden, MA: Polity Press.

Tong, J. (2014). Environmental risks in newspaper coverage: A framing analysis of investigative reports on environmental problems in 10 Chinese newspapers. *Environmental Communication, 8*(3), 345–367. doi:10.1080/17524032.2014.898675

Tong, J. (2016). *Investigative journalism, environmental problems and modernization in China.* Basingstoke: Palgrave Macmillan.

Wang, Z., He, H., & Fan, M. (2014). The ecological civilization debate in China: The role of ecological Marxism and constructive Postmodernism—Beyond the predicament of legislation. *Monthly Review, 66*(6). Retrieved from http://monthlyreview.org/

World Bank. (2014). World development indicators: Environment. Retrieved from http://data.worldbank.org/topic/environment

Zhang, J. Y., & Barr, M. (2013). *Green politics in China: Environmental governance and state-society relations.* London: Pluto Press.

Zhao, Y. (2008). *Communication in China: Political economy, power, and conflict.* Lanham, MD: Rowman & Littlefield.

Zhao, Y. (2014). Communication, crisis, and global power shifts: An introduction. *International Journal of Communication, 8*, 275–300.

7. *'How Mining Made Australia': Populist Nostalgia and the Spectre of Climate Change in the Television Documentary* Dirty Business

PAOLO MAGAGNOLI

'Land, money and power: this is the epic story of mining'. So begins the television documentary *Dirty Business.* Aired on the Australian public network SBS in January 2013, the documentary's three one-hour-long episodes presented a dizzying array of visual sources, including interviews with prominent historians, journalists, and academics; stunning aerial footage of gigantic mining pits and haul trucks in the desert; endlessly snaking trains full of ore; newsreel; newspaper headlines; drawn animations; and rare archival photographs of nineteenth-century gold diggers. The filmmakers—Jacob Hickey and Alex West—worked hard to dramatize the multifaceted history of mining through the grandiose statements of the voice-over, played by stage actor Colin Friels, a bombastic soundtrack, and three-dimensional animations of archival photographs, giving the viewer the almost tactile illusion of a particularly intense bodily encounter with Australia's distant past. The outcome of these visual strategies is a fast-paced, entertaining documentary that addresses a complex and highly controversial subject, although it inadvertently celebrates the industry by emphasizing the awesome visuality of its vast scale and sublime technology.

Dirty Business's reception was largely favourable. The film was praised for its original approach to the history of mining, its important pedagogical value, and its entertaining pace. 'The story, fashioned by writers Alex West and Jacob Hickey', wrote *The Age's* television critic, 'is fascinating and largely convincing: a truly impressive attempt to mine the depths of our history from a genuinely fresh perspective' (Quinn, 2013, p. 2). Likewise, *The Australian* praised the film enthusiastically: 'This three-part series about mining should be mandatory viewing for all Australians', the newspaper asserted. 'This is at once engaging television and an excellent primer on how we went from the gold rush to where we are today' (Cuthbertson, 2013).

Despite its positive reviews, *Dirty Business* made for some bad history. In particular, the story presented by Hickey and West contained some simplifications and omissions (Burnside, 2013). Among these, its failure to consider the environmental effect of mining stands out as the documentary's most striking oversight. For example, it overlooks recent findings relating to the ecological impact of nineteenth-century gold mining. According to historians, gold mining in Victoria caused the destruction of native vegetation, the extinction of many native birds and animals, and water pollution with significant and far-ranging economic costs (Frost, 2013; Garden, 2001, 2005; Harris, 2014). What is more, across the three-hour-long show, the filmmakers completely ignore the role played by the mining sector in exacerbating global warming and the demands of Australian environmentalists to reduce coal and gas extraction in response to the urgency of climate change. Coal and gas exports account in large part for Australia's negative track record on climate change. The nation is one of the biggest exporters of these energy resources and has one of the highest levels of greenhouse pollution per person in the industrialized world. As Peter Christoff has remarked, if the country really wants to respect targets for the reduction of global warming agreed on at the United Nations Climate Change Conference in Paris in 2015, it will have to drastically reduce the extraction of coal and gas. According to Christoff:

> In 2012, Australia emitted some 552 million tonnes (Mt) CO_2. It ranks 12th among the planet's 195-plus nations for its domestic greenhouse gas emissions and contributes some 1.3 per cent of total global emissions. It is 14th for domestic CO_2 emissions alone. Its per capita emissions are among the world's highest. Further, when emissions associated with Australian coal and gas exports are added to its domestic greenhouse emissions, Australia is the source of over 4 per cent of total global emissions. In all, Australia is a major national emitter, a very significant contributor to global warming, and should shoulder part of the additional reduction burden associated with bridging the 'ambition gap' and preserving the 2°C 'guardrail'. (2013, p. 241)

None of this information is provided by the authors of *Dirty Business*. The film's omission of climate change science is both striking and indicative of the marginal place that global warming assumes in the agenda of Australian media (Bacon, 2011, 2013).

History, Memory and Climate Change

Dirty Business conveys ideologically ambivalent messages. On the one hand, it exposes the dark side of the history of mining and some of its negative impacts. The documentary mentions the land dispossession of Torres Strait Islander and Aboriginal peoples by mining companies; the anti-Chinese racism of the White Australia immigration laws in mid-nineteenth century; the violent riots against Yugoslav and Italian migrants in Kalgoorlie at the turn of the century; early miners' appallingly unsafe working conditions; and the crushing of trade unionism. On the other hand, through its spectacular cinematography, the film ends up celebrating the technological power of the industry. The film's ambivalence is further exacerbated by its overarching argument: the centrality of mining for Australian nation-building. In *Dirty Business*, mining is given as one of the reasons for Australian exceptionalism, namely its optimism about the nation's ability to achieve infinite growth and to escape the laws of cyclical economic recession (the country emerged, unscathed, from the 2008 financial crisis). In the introduction to each episode, viewers see spectacular pictures of open-cut mines in the outback accompanied by a voice-over that claims: 'During the last 150 years, mining has been the battleground that has divided and yet forged the nation: the story of mining is Australia's story'. The aerial camera speedily zooms out from a red, canyon-sized hole populated by house-sized trucks. 'Whatever the power and influence of mining really is', the narrator concludes, '150 years of history has proved that its rocky relationship with government will forge the future of Australia for many more years to come'. Hickey and West's script validates the notion that Australia cannot disentangle itself from the development of mining.

The perspective offered by the filmmakers is not unanimously shared by historians. Sarah Burnside, for example, has taken issue against the determinism of *Dirty Business*'s approach to history and its inflated claims. She explains:

> While the program documented some of the negative impacts of mining [...], it never queried the underlying assumption that Australia was and is utterly reliant on the mining industry, for good or ill. There surely could have been some space made for counter-arguments such as those set out in the Australia Institute's 2011 paper 'Mining the truth: the rhetoric and reality of the commodities boom', in which authors David Richardson and Dr Richard Denniss note that in 1951, agriculture 'accounted for just over 30 per cent of Australia's GDP—much bigger than mining has ever been' and argued that the 'rise of the mining industry is neither inexorable nor universally beneficial'. (Burnside, 2013)

In this sense, *Dirty Business* does not offer 'a new and refreshing perspective' on mining history but, instead, a slightly revised version of the dominant view.

The idea that mining is central to the building of Australia is hardly a novelty. Though with different nuances, the argument has been promoted by a significant sector of the culture industry since at least the 1970s (White, 1981). Over the past fifty years, it has also been repeatedly advanced by staunch nationalist and conservative historians such as Professor Geoffrey Blainey, who frequently appears in *Dirty Business.* Blainey is the author of a long series of books that glorify and celebrate the mining industry (Blainey, 1963, 2001, 2009), and he has collaborated with many of BHP Billiton's corporate videos. The argument can also be found in the writing of award-winning journalists and the tabloid press. In the introduction to his 2013 volume, *Boom: The Underground History of Australia, from Gold Rush to GFC*, Malcolm Knox writes: 'This book is an attempt to put up and test a hypothesis that mining is, in a way that has not been properly understood or appreciated, integral to Australians' perception of themselves; that mining is, in a metaphoric sense, woven into the national DNA' (Knox, 2013, p. xii–xiii). Furthermore, the idea that mining represents the 'backbone of the country' is harnessed by the industry to promote and advertise itself; it is used by the Liberal-National Coalition in government to enforce their pro-mining sector policies, and it continues to be a powerful narrative in Australia's political culture. Consider the following statements by prominent political and business leaders:

> Our prosperity rides on the ore and gas and coal carriers steaming the seas to our north, just as surely today as once it rode on the sheep's back. (former Prime Minister Tony Abbott, 2014)

> Mining is a core part of Australian history. Every Australian schoolchild learns at least one thing about Australian history: the Eureka Stockade … Mining is part of our collective history. It has driven so much of the economics and politics of our present and it will continue to play a hugely important role in Australia's future. (Minister for Resources and Northern Australia, Matt Canavan, 2016)

> [Coal mining] is the pride of Australia's past, and the pride of its future. (Peabody Energy Chairman and CEO Gregory H. Boyce, The University of Queensland, 2009)

Likewise, *Dirty Business* never questions the underlying assumption that mining represents the pride of both Australia's past and its future. Instead, the documentary implicitly suggests that, given the country's still-untapped reservoir of mineral resources *and* its central role in the nation's history, extraction will continue to shape the future and destiny of the country.

National historical narratives play an important role in reinforcing denial about global warming. As Kari Mari Noorgard has shown in her ethnographic study of Norwegians' ambivalent attitudes towards global warming

(Noorgard, 2011), stories about national identities can act as 'cultural tool-kits' to contain guilt about the wealth generated by the extraction of fossil fuels (Norway, like Australia, is one of the top producers of oil and natural gas in the world). Noorgard's study shows that the politics of climate change need to take into account narratives about the future as well as stories about the past, because the latter can significantly impact people's attitudes about global warming. Mass media are an important vehicle for the circulation of such 'cultural tool-kits', and documentaries such as *Dirty Business* are important objects in the study of the relation between the media and climate change politics. In this chapter, I argue that the story that mining is central to Australia's national identity is an important 'cultural tool-kit' mobilized by the mineral industry and the political right in Australia to justify the continuation of fossil fuel extraction in the face of global warming.

A Populist Nostalgia

Why is the myth of mining as 'the story of Australia' so ingrained in the country's collective memory? Why is it so powerful today despite widespread recognition that the profits of the last mineral boom have not benefited the majority of its citizens? The story of mining as 'the story of Australia' remains a powerful myth because it is infused with a feel-good nostalgia that appeals to the populist sentiments of a widespread sector of the public. According to this nostalgic approach to history, the extractive industry emerges as the last remnant of a romanticized past. This past is located in the late nineteenth century and early twentieth century—that is to say, the period when the country stopped being a colony and became independent from the British Empire. The outback is the space where these nostalgic narratives usually unfold. A vaguely defined area of the country's vast inland territory, the outback plays a central role in the Australian nationalist imaginary (White, 1981). It is also the area where most mines are placed. The mining sector's public relations and advertisement campaigns exploit the outback imaginary and reframe it as the populist 'heartland' (Taggart, 2000), in other words as the space where the 'real people' live as opposed to the city, or the space of the intellectual and corrupt elite.

A populist nostalgia infuses several passages of the documentary. With its images of heavy haulage trains traversing the red Australian desert and rugged miners, *Dirty Business* builds upon this fascination for the outback. The directors deploy a plethora of archival materials, which, at times, appear excessive and superfluous to the development of the narrative. Their obsession for the archive might be explained as a desire for historical presence that animates much contemporary popular culture. Film theorist Jamie Baron calls this impulse the 'archival affect' (Baron, 2014).

> The desire for presence is the desire for the archive '*affect*', she explains, 'for an awareness of the passage of time and the partiality of its remains, for an embodied experience of confronting what has been lost, and the mortal human condition. We seek to renew this affect again and again in order not only to *know* the past but also to *feel* it'. (Baron, 2014, p. 128)

However, as Baron observes, there are pitfalls in this fascination for the materiality of the archive: 'There is a danger', she points out, 'that accompanies the archive affect insofar as it verges on nostalgia for a romanticized and idealized past and a fetishization of the archive' (Baron, 2014, p. 128). Baron's warning seems valid for Hickey and West's documentary as well. In *Dirty Business*, old black-and-white photographs are animated through digital software that transforms them from bi-dimensional to three-dimensional pictures. As the camera pans over images of old shanty towns and portraits of nineteenth-century gold diggers, we are given the illusion to enter and move across their decrepit sheds and squalid homes. The effect is often uncanny: the workers' ghostly faces and fragile bodies seem to hover towards us. The appeal of this imagery is not factual but purely emotional; it testifies to the desire to resurrect the past and the nostalgic approach that characterizes the film.

Dirty Business often links past and present through editing. Consider a sequence in the second episode when the narrator relates the discovery of gold in Kalgoorie, Western Australia, in the 1890s. Coinciding with a moment of ravaging economic depression, the discovery ignited the hope of a new gold rush and attracted a massive flow of migrants from the east of the country and from Italy and Eastern Europe. The sequence assembles pristine black-and-white photographs of those migrants with footage of contemporary miners. In the photographs, a worker carries a wheelbarrow with his few personal possessions; others are riding camels in the desert of Western Australia, while others stand in front of makeshift tents and drab camping sites. The camera cuts to shots of contemporary fly-in fly-out miners standing in airports and waiting to reach their workplace. In the meantime, adding to the emotional tone of the imagery, a slow, melancholic piano score can be heard in the background. Through evocative passages like this, *Dirty Business* not only pays tribute to early Australians' extraordinary endurance and exceptional character, but it also suggests that today's miners are the heirs of nineteenth-century gold diggers—that is to say, the present embodiment of the most 'authentic' Australians.

The dangers of *Dirty Business*'s nostalgic populism emerge in those moments of the film where mining is celebrated as the heart and soul of Australian democracy. A particularly long section of the documentary focuses on the seminal event of the 1854 Eureka Stockade. This was a rebellion of the goldfield workers in Victoria (known as 'diggers') that emerged in opposition to the government miners' licences. The licences were a simple way for the government to tax diggers. Licence fees had to be paid regardless of whether a digger's claim resulted in any gold. Less successful diggers found it difficult to pay their licence fees and, as consequence, began a violent protest that was crushed by the colonial forces, causing the death of 27 people. The rebellion led to the introduction of the Miner's Right, which established the right to vote and own land for miners. In an attempt to vivify and resurrect this important historical event, the filmmakers re-enact the rebellion through the animation of drawings and sound effects of explosions and screams. More importantly, Hickey and West overlay these animations with statements from interviewee Peter Hartcher, political editor of *The Sydney Morning Herald*, who defines the gold diggers as the 'shock-troops of Australian democracy'. The narrator further endorses this reading of the Eureka Stockade, declaring that 'mining has been a catalyst for Australia to become one of the most democratic and egalitarian places on earth'.

'Nostalgia, like any form of narrative, is always ideological', writes Susan Stewart in her book *On Longing* (1984, p. 23). Elsewhere, she explains: 'Hostile to history and its invisible origins, and yet longing for an impossibly pure context of lived experience at a place of origin, nostalgia wears a distinctly utopian face, a face that turns toward a future-past, a past which has only ideological reality' (Stewart, 1984, p. 23). This definition of nostalgia seems valid for Hickey and West's story of mining. The filmmakers treat the Eureka rebellion as a watershed moment in Australian and mining history—a moment of origin whose meaning is unequivocal and clearly settled. In fact, since the 1850s, the event has been written and re-written so many times across Australian history and culture, and from so many different points of view, that it is no longer possible to distinguish between fact and fiction. As historian Chris Healy has pointed out, the rebellion has achieved the status of legend. 'Eureka', Healy argues, 'has remained one of the few symbolic traces of a 'golden age' when the Australian colonies were transformed by migration and wealth; a time that, for those who lived through the 1850s, was imagined as both a deformation and a new beginning' (Healy, 1997, p. 156). Eureka has been appropriated from very different political movements and parties and for very different ideologies and purposes throughout the last century. In the 1950s, it was used as a symbol of egalitarianism by Australian Maoists, and in the 1970s, it was also appropriated by the political right to support tax evasion. The fact that the meaning of Eureka is pliable to so many uses and interpretations makes it a particularly unstable and ambiguous sign.

Rather than accounting for the complexity of this event, Hickey and West do not propose a new, alternative reading of the Eureka stockade. They portray the famous rebellion as a foundational moment in the building of the Australian democracy, and yet they safely refrain from explaining in detail the radical ideas and demands made by the diggers. More importantly, they do not ask whether—today, as in the past—the mining industry's immense wealth was and truly is shared in a democratic and egalitarian way; that is to say, Hickey and West do not investigate whether or not the mining sector has *actually* favoured the egalitarianism advocated by the Eureka rebels. In other words, the positive utopian content of the famous historical event could have been used to gauge the present and identify the shortcomings of the current political system. The directors refuse to clearly spell out the distinction between past and present and prefer to rely on the charming but elusive meanings of the archival photographs. In the process, *Dirty Business* seems to confuse the 'mining industry'—as a specific profit-driven organization owned by national and international shareholders—with the figure of 'the miner' as a metonym for the concept of 'the people'.

The filmmakers' confusion of mining with democracy is symptomatic of the populist mood underpinning the documentary. Populism is a slippery and fuzzy concept, and, as an ideology, traverses both left and right sides of the political spectrum. According to political scientist Cas Mudde, populism is a 'thin-centred' ideology; in other words, it is a malleable and sometimes incoherent set of ideas that can be adopted by movements with very different political agendas. Mudde has argued that, however vague the concept is, we can still identify some ideological elements at the heart of every populist movement. He elaborates the term as 'a thin-centred ideology that considers society to be ultimately separated into two homogeneous and antagonistic groups, 'the pure people' versus 'the corrupt elite', and which argues that politics should be an expression of the *volonté générale* (general will) of the people' (Mudde, 2007, p. 23). The problem with populism, Mudde points out, is that it turns the figure of the people into some monolithic and pure category:

> The main bad is that populism is a monist and moralist ideology, which denies the existence of divisions of interests and opinions within 'the people' and rejects the legitimacy of political opponents. As the populists are the *vox populi*, ie the voice of all the people, anyone with a different view speaks for 'special interests', ie the elite. (Mudde, 2015)

By glorifying the figure of the miner and presenting it as the embodiment of the nation's libertarian and democratic impulse, Hickey and West implicitly reassert the dichotomy between the 'pure people' versus 'the corrupt elite' that underpins populist ideologies. While the filmmakers are also careful to point out that today's protesters are mining billionaires such as Gina Rinehart, they nevertheless tend to indulge in the romanticisation of the miner as a symbol for the 'pure people', and inadvertently end up reinforcing populist attitudes.

The documentary's representational strategies recall other depictions of the industry that have been highly successful in Australia in recent years and have reached mainstream status: the working-class miner as populist hero is a recurring trope of mineral industry advertisement campaigns and corporate communication. Take, for example, the series of commercials entitled the 'This is Our Story' campaign, launched by the Mineral Council of Australia in 2011. Each commercial in the series is shot as a spontaneous documentary of the everyday life of contemporary miners, who are shown smiling and engaging in playful acts of camaraderie. The creators of the campaign exploited the authenticity of the miner to suggest that mining companies are equivalent to the people. As Shelley Marshall explains:

> The strategy creates a sense that the mining industry is integrated into Australian social fabric. It counters the idea that the mining industry is an 'enclave' industry and builds an emotional resonance between viewers and the companies. By invoking the 'Australianness' of the individuals who work in the mines, we form intimate and direct relationships with people we identify with, and mining companies lay an implicit claim to the mineral assets that lie under Australian soil. (Marshall, 2012)

Another example of the deceptive populism that often underpins narratives about mining in the mainstream media is the feature film *Red Dog* (Stenders, 2011). A great commercial success (it was the highest grossing domestic film of that year, and sits eighth on Screen Australia's 2012 list of top Australian films at the box office), the film exemplifies another stereotypical representation of the miner as the 'authentic' voice of the people. *Red Dog* is set in the red Pilbara Desert in the 1970s, and it relates the life of a community of likeable and roguish miners who work for Hamersley Iron (a wholly-owned subsidiary of mining giant Rio Tinto). As in *Dirty Business*, the cinematography is lush: the camera sweeps across the red Pilbara landscape, depicting mines as part of the raw beauty of the landscape. Director Stenders has candidly declared that the film it is a 'celebration' of the birth of the mining industry. 'When you are up there you realise that this is the heartbeat of Australia. It's very sobering to see the infrastructure and scale of it' (Stenders as cited in Barkham, 2012). However, other critics have denounced its lack of Indigenous characters and its sentimentalising of the lucrative, but environmentally destructive, mining industry (Burnside, 2011). Although Hickey and West are more ambivalent in their portrayal of mining, in *Dirty Business* they nevertheless adopt similar representational strategies and themes to those of *Red Dog* and *This is Our Story*.

Who Controls the Past Controls the Future

Climate change is the elephant in the room in the 2013 television documentary *Dirty Business*. The three-hour-long film proposes a strictly anthropocentric view of the history of Australia, which suggests that the country's past and future cannot be disentangled from the history of the extractive sector. The documentary nevertheless ignores the debate around climate change in Australia and environmentalists' calls for fossil fuel divestment. It eschews the discussion of the ecological impact of the industry in favour of a nostalgic account that tends to monumentalize mining. This nostalgic approach has strong populist connotations: the film conflates the figure of the miner with the figure of the people and suggests that Australian democracy and equality are deeply indebted to the industry. *Dirty Business* suggests that the extraction and consumption of natural resources is a patriotic duty and a right inscribed in the nation's past; in this way, it ultimately defuses the threat of climate change by treating it as a reality that never really happened.

Importantly, the documentary's populist tendency recalls some of the advertising strategies deployed by the mining sector to promote itself, which are exemplified by films such as *Red Dog* or the commercials belonging to the 'This is Our Story' campaign. *Dirty Business*'s conclusion—that 'mining will continue to shape Australia for years to come'—evokes a fatalistic understanding of history that is clearly problematic in light of the dangers of climate change. Insofar as it endorses the myth that 'mining has made Australia', *Dirty Business* upholds what Guy Pearse has called the country's unshakable 'quarry vision', that is to say the idea that Australia's past and future prosperity depends on 'what we dig, drill and smelt for the world' (Pearse, 2009, p. 2). 'Not for a moment', Pearse argues, '[in Australia] do we foresee an inevitable collision between climate change and the quarry. Not for a moment do we pause and try to imagine a quarry without coal' (Pearse, 2009, p. 3). The story of *Dirty Business* exemplifies the ways in which historical knowledge shapes not only national consciousness but also climate change politics and alternative visions about the future.

Commenting on the tremendous power of cinema to construct popular memory, Michel Foucault remarked that, 'if one controls people's memory, one controls their dynamism' and one 'impose[s] a framework to interpret the present' (Foucault, 1989, p. 130). For Foucault, power forces past attempts to produce social change into oblivion in order to defuse present struggles. Collective memory, by this formulation, can be dangerous because galvanizes transformative and revolutionary impulses. Yet, as the case of *Dirty Business* demonstrates, collective memory can also block positive utopian thinking. Put another way, it can undermine the imagination of alternative and 'greener' economies and modes of production. Climate change advocacy should not rely solely on the collection and dissemination of more and better facts but instead look at the ways in which collective memory and national myths engage people at the level of emotion.

References

Abbott, T. (2014). *Address to the minerals week, annual minerals industry parliamentary dinner*. Retrieved November 12, 2016, from http://pmtranscripts.pmc.gov.au/release/transcript-23528

Bacon, W. (2011). *A sceptical climate. Media coverage of climate change in Australia*. Sydney: The University of Technology Sydney.

Bacon, W. (2013). *Climate science in Australian newspaper*. Sydney: The University of Technology Sydney.

Barkham, P. (2012, February 9). Red Dog: An audience with Australia's best friend. *The Guardian*. Retrieved January 12, 2017, from www.guardian.co.uk/film/2012/feb/09/red-dog-australia-bestfriend.

Baron, J. (2014). *The archive effect: Found footage and the audiovisual experience of history*. London; New York, NY: Routledge.

Blainey, G. (1963). *The rush that never ended: A history of Australian mining*. Carlton: Melbourne University Press.

Blainey, G. (2001). *This land is all horizons: Australian fears and visions*; 2001. Sydney: Australian Broadcasting Corporation.

Blainey, G. (2009). *A shorter history of Australia*. North Sydney: Random House Australia.

Burnside, S. (2011, August 16). Red Dog whitewashes the Pilbara. *New Matilda*. Retrieved January 12, 2017, from www.newmatilda.com/2011/08/16/red-dog-whitewashes-pilbara

Burnside, S. (2013, January 24). Dirty business. *Onlineopinion*. Retrieved December 12, 2017, from http://www.onlineopinion.com.au/view.asp?article=14609.

Canavan, M. (2016, August 22). Mining is central to Australian history and has a strong future. *The Australian*. Retrieved January 12, 2017, from http://www.theaustralian.

com.au/opinion/mining-is-central-to-australian-history-and-has-a-strong-future/news-story/451fc21ef62b338467bfe301b17570a9

Christoff, P. (Ed.). (2013). *Four degrees of global warming: Australia in a hot world.* Abingdon: Taylor and Francis.

Cuthbertson, I. (2013, January 5). Dirty business: Story of greed, glory, power and politics. *The Australian online.* Retrieved January 24, 2017, from http://www.theaustralian.com.au/arts/review/dirty-business-story-of-greed-glory-power-and-politics/news-story/19363cc682cec92512469be662fafef8

Foucault, M. (1989). Film and popular memory. In Sylvère Lotringere (Ed.), *Foucault live. collected interviews, 1961–1984.* New York, NY: Semiotext(e), 122–132

Frost, W. (2013). The environmental impacts of the victorian gold rushes: Miners' accounts during the first five years. *Australian Economic History Review, 53*(1), 72–90.

Garden, D. (2001). Catalyst or cataclysm? Gold mining and the environment. *Victorian Historical Journal, 72*(1–2), 28–44.

Garden, D. (2005). *Australia, New Zealand, and the Pacific: An environmental history.* Santa Barbara, CA: ABC Clio.

Harris, E. (2014). Property rights regimes and their environmental impacts. In S. Ville (Ed.), *The Cambridge economic history of Australia.* Cambridge: Cambridge University Press.

Healy, C. (1997). From the ruins of colonialism: History as social memory. Cambridge; Melbourne: Cambridge University Press.

Knox, M. (2013). *Boom. The underground history of Australia, from gold rush to GFC.* Melbourne: Viking.

Marshall, S. (2012, October 11). Hearts, minds and hip pockets: How the resources industry aims to win over ordinary Australians. *The Conversation.* Retrieved February 7, 2017, from www.theconversation.com/hearts-minds-and-hip-pockets-how-the-resources-industry-aims-to-win-over-ordinary-australians-10028

Mudde, C. (2007). *Populist radical right parties in Europe.* Cambridge: Cambridge University Press.

Mudde, C. (2015, February 17). The problem with populism. *The Guardian.* Retrieved March 7, 2017, from www.theguardian.com/commentisfree/2015/feb/17/problem-populism-syriza-podemos-dark-side-europe

Noorgard, K. M. (2011). *Living in Denial: Climate change, emotions, and everyday life.* Cambridge, MA: MIT Press.

Pearse, G. (2009). Quarry vision: Coal, climate change and the end of the resources boom. *Quarterly Essay, 33*, 1–122.

Pearse, G., McKnight, D., &, Burton, B. (2013). *Big coal. Australia's dirtiest habits.* Sydney: New South Publishing.

Quinn, K. (2013, January 10). Dirty business: How mining made Australia Sunday, SBS One, 8.30pm. *The Age*, 10.

Stenders, K. (2011). *Red Dog.* Endymion Films.

Stewart, S. (1984). On longing: Narratives of the Miniature, the Gigantic, the Souvenir, the Collection. Durham and London: Duke University Press.

Taggart, P. (2000). *Populism*. Buckingham; Philadelphia, PA: Open University Press.

The University of Queensland. (2009). *Coal in the commonwealth: Study points to the future of coal in Australia*. Retrieved December 12, 2016, from www.uq.edu.au/news/article/2009/10/coal-commonwealth-study-points-future-of-coal-australia

White, R. (1981). *Inventing Australia: Images and identity: 1688–1980*. Sydney; Boston, MA: Allen & Unwin.

8. *See It Before It's Too Late? Last-Chance Travel Lists and Climate Change*

Lyn McGaurr and Libby Lester

Introduction

> Any time I've pitched a story because of the environment, or using the environment as a hook, it's always met a lukewarm response. The only time it doesn't is when it's the Maldives conundrum, and it's 'Go now, before it's gone'. (US travel journalist, pers. comm. 2015)

In an era of rapidly increasing human population and global travel, 'last-chance tourism' is attracting considerable media attention (Dawson, Lemelin, Stewart, & Taillon, 2015). This phenomenon is defined by scholars as commercial tourism that exploits 'vanishing landscapes or icescapes, and/or disappearing natural and/or social heritage' (Lemlin, Dawson, Stewart, Maher, & Lueck, 2010, p. 478). Many last-chance destinations and experiences are ones at risk from climate change—polar bear viewing in Canada, the ice-capped peak of Mount Kiliminjaro in Tanzania, the glaciers of Greenland, and Australia's Great Barrier Reef (e.g. Frew, 2008, 2012; Lemelin, Dawson, Stewart, Maher, & Lueck, 2010; Piggott-McKellar & McNamara, 2016). Tourism scholars frequently cite published lists of places to go before it's too late as evidence that media on the subject is proliferating (Coglan, 2012; Dawson et al., 2015; Frew, 2008; Lemelin et al., 2010; Lemelin, Stewart, & Dawson, 2012; Piggott-McKellar & McNamara, 2016). A 2015 survey of visitors to the Great Barrier Reef found that most had developed their perceptions of it from the media and nearly 70% were motivated to see the Reef before it was too late (Piggott-McKellar

& McNamara, 2016). The study identified high levels of concern about coral bleaching and climate change but produced little evidence that tourists appreciated the contribution to damaging global warming made by greenhouse gases emitted in the course of their own airline travel (Piggott-McKellar & McNamara, 2016). In a subsequent interview about the research, one of the study's authors (Piggott-McKellar in Hill, 2016, paras 8–9) said media had a responsibility to alert tourists to 'the potential ecological damage involved in visiting those areas' and the power 'to inform the general public of the threats facing tourist destinations'.

The way media frame the issue of climate change can influence how audiences think about it (Boykoff, 2007; Nisbet, 2009). Hansen and Machin (2008, p. 779) advocate media representations that 'locate and connect [environmental] issues in actual concrete processes such as global capitalism and consumerism' (Hansen & Machin, 2008, p. 779), while Nisbet (2009) notes that a frame promoting responsible stewardship of the Earth may empower people to take a greater interest in climate policy and decision-making. Audiences of media about globally significant natural attractions such as World Heritage sites are popular targets of antagonists in disputes about the performance of governments responsible for ensuring the 'identification, protection, conservation, presentation and transmission [of the sites] to future generations' (UNESCO, 1972). Tourism scholars often express concern about the contradictions inherent in promoting a region for its vulnerability while encouraging tourism that may contribute to further degradation, but they pay far less attention to the potential for last-chance travel media to elucidate the complex politics of climate change for a transnational audience (McGaurr, 2015). To address this gap, we analyse 12 last-chance travel lists and assess their contribution to transnational climate-change engagement and debate. We begin by asking the following questions of our data:

- Is climate change prominent in last-chance travel lists?
- Do last-chance travel lists identify the contribution of tourism to greenhouse gas emissions?
- How do images and linguistic text in these lists interact to convey information about climate change?

We then present a case study of the Great Barrier Reef that draws on the same sample to investigate whether last-chance travel lists position readers as members of transnational publics.

Background

Tourism, the Media and the Public Sphere

Climate change is a global problem with local consequences but also local causes. Mines in Australia produce coal for domestic use and export. Like greenhouse gases from aircraft, emissions from the burning of Australian coal domestically and internationally fuel global warming that contributes to rising ocean temperatures capable of bleaching coral in reefs around the world (Tarte, Hart, Hughes, & Hussey, 2017). A plan by Indian company Adani to create one of the planet's largest coal mines in the Australian state of Queensland has been the source of intense environmental conflict since 2010. Opponents of the Carmichael mine have identified the Great Barrier Reef World Heritage Area as a potential victim of the mine via climate change (Australian Conservation Foundation, n.d.; Australian Marine Conservation Society, n.d.; Fight for Our Reef, 2015, 2017; GetUP! n.d.; Greenpeace Australia Pacific, 2016; Stop Adani, n.d.), as has a report to the World Heritage Committee by the Great Barrier Reef Independent Review Group (Tarte et al., 2017). Environmental organisations claim this puts at risk the Australian and Queensland governments' reputation as responsible stewards of a global environmental asset.

Tourism is a 'booster rocket' for reputation management favourable to broader economic aspirations (Anholt, 2007, p. 88), and the media are regarded by place-branding proponents as enthusiastic allies (Anholt, 2010, p. 142). Audiences of media about tourism destinations are not always or only tourists. Whether ardent jetsetters or armchair travelers, they are likely to have additional occupations, personas, business interests, and consumer needs. Flattering coverage of a country or region may incline them to look favourably on its exports and investment opportunities. Governments and jurisdictional tourism offices frequently offer international news, business and travel journalists free transport and accommodation in the hope of positive publicity for policies, products, and/or places (Mackellar & Fenton, 2000; Taylor, 2015). In 2015, for example, the Australian government invited 'journalists from around the world on an all-expenses paid trip to the Great Barrier Reef as part of intensifying diplomatic efforts to try to prevent the reef being placed on the Unesco world heritage committee's "in-danger" list' (Taylor, 2015). The most prestigious media organisations ban such 'freebies', but others have more flexible approaches, such as allowing journalists to accept hosted travel as long as they declare it in their stories. Unlike news journalists, travel media have a reputation for producing purely celebratory copy, and this can help shield national or regional governments from reputational damage from environmental conflict. Due to the necessity and expense of travelling to experience the destinations they cover, many travel journalists consider complimentary visits essential to their jobs (Hanusch, 2012). Tourism public relations practitioners believe recipients of hosted travel are unlikely to 'bite the hand that feeds them' (Dore & Crouch, 2003, p. 148), and this view is partially supported by the results of a survey of Australian travel journalists in which half the respondents agreed that 'free travel does influence reporting to some extent' (Hanusch, 2012, p. 679). Nevertheless, the advantages of tourism communication as a form of place-branding and reputation management can become a liability for governments if travel media behave uncharacteristically by including political observations in their texts. Environmental conflict is still rare in travel media, but some journalists and editors have been prepared to include it on occasion (McGaurr, 2015). Reasons provided by travel journalists include the hope that it will encourage decision-makers to act in the best interests of the destination and global tourism, and an expectation that their audiences will be interested in any similarities between debates in distant places and disputes in their own backyards (McGaurr, 2015). Pro-development governments sensitive to environmental conflict in media about natural attractions claim it damages the tourism industry by discouraging visitors. However, they are also likely to be concerned that it will induce international investors and consumers of products other than tourism to view the country or region less favourably (see Lester, 2016), or persuade foreign publics to support environmental campaigns against contentious developments. Such campaigns attempt to call into being a public sphere in which 'state actors may choose to side with civil society actors in other states to obtain changes in the policies in the latter' (Köhler, 1998, p. 233). For Lester (2016), environmental politics are already being impacted by a perception that transnational publics are emerging, even if the technicalities of their existence have yet to be empirically confirmed. She advocates following 'the generation, circulation and contestation of the symbolic and the spectacular' to reveal 'at least some of the conditions under which meaning-making, attributing responsibility and collective decision-making are taking place transnationally' (Lester, p. 798).

Last-Chance Travel Lists

Two features of last-chance travel lists are central to their transnational communicative potential: the celebritisation of lists (McGaurr, 2015) and the 'spectacle-isation' of nature (Goodman, Littler, Brockington, & Boykoff, 2016). Contemporary online publishers are engaged in fierce competition for social media 'shares' to secure advertising revenue. For these individuals and businesses, lists have become lucrative 'scaffolding for stories' (Shepherd in Fernando, 2013) because they are popular with audiences and perfectly structured for digital transmission on mobile devices. This means online travel lists are now 'cognitively and digitally networked into the hierarchies that populate twenty-first-century media, ceaselessly distributing and redistributing salience' (McGaurr, 2015, p. 166). Their ability to circulate so far and so fast makes them the transnational celebrities of twenty-first-century travel media. When an online travel list labels destinations 'places to see before it's too late', it creates a transnational environmental spectacle, because it constructs the devastation of many of the world's 'natural wonders' as immanent and dramatic. It is in their capacity as transnational environmental spectacles that last-chance travel lists have the potential to contribute to meaning-making about climate change and challenge the public relations advantages that traditional travel media afford pro-development governments.

Method

Our empirical work begins with a content analysis of last-chance travel lists, narrowing first to closer examination of those entries in the lists that identify climate change as a problem and then to entries about the Great Barrier Reef. As photography is intrinsic both to the spectacular and to travel lists, our analysis pays particular attention to the kinds of photographs selected for the lists, and the ways photographs and linguistic texts interact.

In May 2016 we searched 'see it before it's too late' on Google and created a sample of all relevant travel lists appearing in the first 10 pages of returns. This yielded 12 lists containing a total of 142 items. All but one list had a headline that was a version of 'before it's too late', 'before they're changed forever' or 'before they're gone'. The other list referred to 'before they disappear' in the subheading below its headline. Although some of the lists did not provide a publication date, we estimate that they were all published between 2009 and 2016 inclusive. List publishers appeared in our Google returns in the following order:

- *Jetsetter* (part of the Tripadvisor stable)
- *EscapeHere* (part of the Canadian media company Concourse)
- *LifeHack* (lifestyle blog)
- *Allianz* (European insurance company)
- *Afrizap* (African online content aggregator)
- *Escape* (online travel section for News Corp's Australian tabloid mastheads)

- *CNN.com* (US news site)
- *Inhabitat* (Californian design blog)
- *Lush Palm* (US surf travel blog)
- *DestinationSeeker* (US travel blog)
- *Destination Travel* (on US news site NBCNews.com)
- *Fodor's Travel* (international travel-guide publisher now owned by Random House).

The first author was the only coder. Each list began with a general introduction. This was followed by individual entries for each destination or activity included in the list. List introductions and individual entries were coded for explicit and implicit references to climate change. To be categorised as explicit, the text needed to include the terms "climate change" or "global warming". Introductions and entries were also coded for whether or not they mentioned tourism as a problem and whether they referred to tourism as contributing to climate change. Images were coded according to whether or not they might be interpreted as referencing climate change as a problem (see below for limitations to this coding category). As the amount of text per item was small and sometimes served as a caption, in the few instances in which separate captions were present they were treated as part of the linguistic text. Two of the lists contained the same entries and entry texts, but we included both these lists in our sample because the introductions to each list and many of the images were different.

Broad Coding Results

Linguistic Texts

Although most last-chance tourism scholarship to date has focused on the polar regions (Piggott-McKellar & McNamara, 2016), the lists in our study included a surprisingly wide range of destinations—a total of 63 across the 12 lists. Only five places appeared in at least half the lists, while 34 appeared in only one list (Table 8.1). Destinations that appeared in most lists were the Great Barrier Reef (9), Venice (8), the Dead Sea (7), the Taj Mahal (7), and Mount Kilimanjaro (6). Antarctica, the Maldives, the Everglades, and the Congo Basin all appeared in five lists. Except in the case of the Taj Mahal, climate change was mentioned explicitly or implicitly in at least one of the entries for each of these nine destinations. Of the 12 list introductions, five referred to climate change explicitly (Table 8.1). In one of these five introductions, climate change was the only problem raised, implying that climate change was a problem for each of the subsequent destinations or activities included in that list. Two introductions mentioned climate change only implicitly, and one of these (Mountain, 2016) was the only introduction to make a connection between tourism and climate change (see below). The other list introduction that referred to climate change implicitly merely noted that 'we live on a planet that's constantly changing because of people and climate alike' (Tara, 2015).

Overall, climate change was mentioned implicitly or explicitly in 64 entries for a total of 26 different destinations (Table 8.3), and in 27 of these entries the phrase 'climate change' or 'global warming' appeared in the text (Table 8.2). In total, seven list introductions and 47 individual entries referred to tourism as a problem (Tables 8.1 and 8.2), but the only linguistic text in the entire study that linked tourism to climate change was the introduction that referred to long-haul flights as a problem but did not mention climate change explicitly (Mountain, 2016). Moreover, any advocacy for behavioural change this single reference might have implied was undercut by the fact that all the entries in the list that followed were represented entirely positively as must-see holiday destinations. One entry about Kilimanjaro did not identify tourism as a problem but did refer to climate change as a problem for the tourism industry. One entry about Antarctica negated its own educative potential with the carelessly worded observation that 'the biggest threat to the ice cap is warming temperatures, not humans' (Yaccato, circa 2015). This entry did not link tourism to greenhouse gas emissions but focused on the impact of climate change on Antarctic tourism by reporting that an Antarctica made vulnerable by climate change would not be able to support as many visitors as it might otherwise.

Table 8.1: Climate change and/or tourism referenced as problems in list introductions.

Total number of lists	12
Entire list about climate change	1
Headline mentions climate change (explicit)	0
Introductory text mentions climate change (explicit)	5
Introductory text mentions climate change (implicit)	2
Introductory text mentions tourism as a problem	7
Introductory text indicates tourism contributes to climate change	1

Table 8.2: Climate change and/or tourism referenced as problems in individual entries.

Number of entries across all 12 lists	142
Number of different places in at least 1 list	63
Entries in at least half the lists	5
Entries in only 1 list	34
Entries mentioning climate change (explicit or implicit)	64
Entries mentioning climate change (explicit)	27
Entries mentioning tourism as a problem	47
Entries mentioning climate change (explicit or implicit) and tourism as problems	6
Entries indicating tourism contributes to climate change	0

Images

The first author coded all images accompanying items referencing climate change for the possibility that they illustrated climate-change impacts. The limitations of this form of coding are obvious and unavoidable but we believe even the subjective coding we present here provides insights that become more valuable when the linguistic texts and images are considered together (see following section). Whatever else they may be, travel lists are selling tools for the tourism industry (McGaurr, 2015). For this reason, the coder paid attention to the aesthetic appeal of the images. If an image was aesthetically unappealing she was more likely to conclude that it conveyed concern over climate-change impacts. If the image was appealing but included anomalous elements (very little snow on the peak of Kilimanjaro, for example) she was likely to conclude it illustrated an effect of climate change. Images of ice, snow, and glaciers were particularly difficult to code in the absence of before-and-after shots—a means by which environmental groups traditionally attempt to make the effects of climate change visible (Doyle, 2007). Ultimately, she coded all three images of the Athabasca Glacier as possibly illustrative of climate change because the images foregrounded exposed earth, which could be interpreted as evidence of the glacier receding. She did not code four of the five images of Antarctica as possibly illustrating climate-change impacts because they were conventionally attractive and did not contain elements such as ice crashing into the sea. However, she coded the fifth image as possibly illustrating climate change because although it was attractive it contained what appeared to be a deep fissure in the continental ice filled with water. Using imprecise measures such as these, she estimated 19 of the 63 images coded might possibly be considered illustrative of climate change impacts when viewed without reference to captions or text other than the name of the destination (Table 8.3). The predominance of images of apparently unspoiled destinations in the lists in our sample accords with what might be expected if last-chance travel lists are regarded as forms of destination publicity promoting a natural product (see, for example, Linder, 2006).

Table 8.3: Images accompanying entries that mention climate change explicitly or implicitly.

Place	**Number of lists**	**Climate change**	**Images possibly illustrating climate change impacts**
Great Barrier Reef	9	8	1
Venice	8	7	0
Dead Sea	7	1	0
Mount Kilimanjaro	6	6	3
Maldives	5	5	3
Everglades	5	1	0
Congo Basin	5	2	0
Antarctica	5	5	1
Seychelles	4	4	1
Alps	4	4	2
Madagascar	4	1	1
Athabasca Glacier	3	3	3
Glacier National Park	3	3	0
Tibet and Himalayas	2	1	1
Mendocino Wine Country	1	1	0
Alaska	1	1	0
Amazon Rainforest	1	1	0
Arctic	2	2	2
Beaches of Oahu, Hawaii	1	1	0
Central America's Cloud Forest	1	1	No image
Chan Chan Archaeological Zone	1	1	0
Jakata	1	1	0
Magdalen Island	1	1	1
Na Pali Coast	1	1	0
Patagonia	1	1	0
South Australia	1	1	0
Total		**64**	**19**

Images and Text

The difficulty we experienced in our visual coding is a function of the polysemy of the image (Barthes, 1977). Over many years, environmental campaign images of unspoilt nature and wildlife have been 'aesthetically [inscribed] into the romantic tradition of landscape images' (Hansen & Machin, 2013, p. 156; see also Doyle, 2007), making them available for appropriation by destination marketing campaigns (Lester, 2007; McGaurr, 2015, 2016). But the opposite has also occurred: beautiful images already associated with particular tourism destinations have been put to the service of environmental campaigns as examples of what might be lost both ecologically and economically (see, for example, Australian Conservation Foundation, n.d.; GetUp!, n.d.; Greenpeace Australia Pacific, 2012). However, it takes time for images to be loaded with meaning to such an extent that they come to signify something as visually illusive as climate change (Hansen & Machin, 2013). Images may have the ability to speak for themselves but in particular contexts they remain '"co-constructors" of environmental narratives that, in combination, convey complex and multidimensional messages to media consumers' (DiFrancesco & Young, 2010, p. 520). When charismatic images of unspoiled nature interact harmoniously with a linguistic text in ways that signify what is at stake, the image is working to reinforce the text through the emotional and cultural resonance of its subject matter (DiFrancesco & Young, 2010). Indeed, there is evidence to suggest that although images that depict the worrying consequences of climate change have strong salience with viewers, they tend to make them feel disempowered and do not promote self-efficacy (O'Neill & Nicholas-Cole, 2008; O'Neill, Boykoff, Niemeyer, & Day, 2013). In our study, when we considered the interaction of images and texts in our sample of list items referencing climate change, beautiful photographs of unspoiled nature simultaneously signified what was to be lost and what must therefore be consumed in haste. As Hansen and Machin explain:

> In the case of advertising and marketing, it is clear that certain kinds of representations are favored that allow products and services to be loaded with a moral sense of connecting with nature and the environment. In all cases, representations appear to favor individual responses to environmental problems rather than those that call for major structural changes in terms of the way in which we organize our societies and the resource greedy nature of capitalism. (Hansen & Machin, 2013, p. 157)

In last-chance travel lists, even the individual responses they advocate are circumscribed: sustainable practices on the ground are promoted, but arguing against long-haul travel is rare. This is hardly surprising, given that encouraging such a change in consumer behaviour would undermine the global visitor economy that travel media inhabit.

Great Barrier Reef Case Study

The 2,500 kilometre long Great Barrier Reef is the largest coral reef in the world and Australia's premier tourism attraction. In 2015 it hosted 2.4 million tourists from Australia and 1.1 million from overseas (Swann & Campbell, 2016). That year, tourists spent $3.3 billion in the area, supporting the jobs of nearly 40,000 people (Swann & Campbell, 2016). The reef is under threat on multiple fronts, including tourism, pesticides, soil run-off and climate change (Lester, 2016). Early in 2016, the reef's worst coral bleaching on record resulted in extensive media coverage in Australia and internationally. Global Factivia searches by the Australia Institute identified 2,887 media articles containing references to the Great Barrier Reef and coral bleaching published between January 1 and June 1, 2016, compared to just 90 stories for the same period the previous year (Swan & Campbell, 2016). A subsequent bleaching event in 2017 prompted the Independent Expert Panel advising the Australian government on its Reef 2050 Long-Term Sustainability Plan to urge the government to revise the plan to 'include climate change adaptation and mitigation actions' (Independent Expert Panel, 2017):

> The Panel were united in their concern about the seriousness of the impacts facing the Reef and concluded that coral bleaching since early 2016 has changed the Reef fundamentally. There is great concern about the future of the Reef, and the communities and businesses that depend on it, but hope still remains for maintaining ecological function over the coming decades [...] The Panel considers that action to reduce emissions of greenhouse gases must be central to the response. (Independent Expert Panel, 2017)

Studies of tourism on the Great Barrier Reef suggest some people will be motivated to visit the Reef if they believe it may be their last chance to see it before it deteriorates, while others will be dissuaded by news of degradation. Although Piggott-McKellar and McNamara's (2016) survey mentioned above of 235 domestic and international tourists visiting the Reef found a large majority were motivated by a desire to see it before it was too late, a much larger survey of more than 4,400 people conducted after the 2016 bleaching found very significant numbers would travel elsewhere in Australia if bleaching on the Reef continued (Swann & Campbell, 2016).

The Australian Government's sensitivity to negative publicity about the Great Barrier Reef is related in large part to the long-running campaign by the environmental movement against the proposed Adani coal mine (see 'Background' section above). In addition, the Reef has been called into service in broader campaigns lobbying the Australian Government to withdraw its support for coal mining generally and set more ambitious targets for greenhouse gas reductions. The Reef was inscribed on the World Heritage List in 1981, but at the time of writing it remained at risk of being listed as 'in danger' by the World Heritage Committee (Miles in Gregory, 2016). Even though the Australian Government publicly acknowledges that climate change is a significant threat to the Reef, its extreme sensitivity to the issue was evident in 2016 when it pressured UNESCO to omit the attraction from a publication entitled *World Heritage and Tourism in a Changing Environment* (Markham, Osipova, Lafrenz Samuels, & Caldas, 2016).

The Great Barrier Reef in Last-Chance Travel Lists

The Great Barrier Reef appeared in nine of the 12 lists we analysed for this study. None of the nine was published after the 2016 coral bleaching event. The Reef's highest rank in any list was first of 10, while its next highest was second of 25. Its lowest rank was eighth of 10. Eight entries referred to climate change as a threat to the reef. In one case, the dual threats of climate change and tourism were highlighted, but tourism was not identified as contributing to climate change. In one entry, tourism was presented as the major threat to the Reef and there was no reference to climate change. Interestingly, four entries referred to damage to the reef that would occur if new coal mines went ahead. In three cases the damage identified was that which would occur from a port development proposed to service the mines. Only in one entry were the mines' contributions to climate change via coal production acknowledged, and here that damage was identified as a risk to tourism as well as the Reef:

> More recent threats are damage caused by the development of Australian ports to export coal to China, thereby contributing to more CO2, hence more damage to the reef and its $4 billion in tourist income. (Yaccato, circa 2015)

The introduction to the *EscapeHere* list in which this entry appeared made some of the 'contradictions' of last-chance tourism explicit and suggested readers might 'solve that moral quandary by designing am [sic.] environmentally sensitive visit. Or contribute to conservancy groups that are fighting to save them' (Yaccato, circa 2015). *EscapeHere*'s individual Reef entry linking proposed Queensland coal developments to climate change did not urge people to visit the Reef but it also did not refer to the greenhouse gas emissions of long-haul flights. By noting that coral reefs in Belize are also suffering from coral bleaching, this list implicitly linked the greenhouse gas contribution of Australian coal with damage to the reefs of other nations.

Of the images associated with entries that mentioned climate change, including the *EscapeHere* entry, two were of coral and colourful fish, and each of these images was repeated in one other unrelated list. Only one image could be interpreted as possibly illustrating climate change impacts, and this was a photograph of a turtle swimming above colourless coral. Almost all of the subheadings that accompanied entries in this list highlighted risks to the destinations concerned. Thus, the image of the turtle was firmly anchored to risk through the subheading 'Much grief for the Great Barrier Reef—Australia' (Destination Seeker, circa 2014–2015). The remaining images of the Great Barrier Reef in our sample were three different aerial shots. One of these illustrated a list entry that did not reference climate change at all, while the other two accompanied the *Jetsetter* and *Escape* entries, which contained identical linguistic texts. Images of the Reef as more than a tropical fish tank—as a massive living entity that clearly extends well beyond the frame—signify its consequence to the Earth's ecological processes and perhaps its simultaneous vulnerability to global environmental change. Aerial images of the Reef are also prominent in activist and news representations of environmental conflict (see, for example, Greenpeace Australia Pacific, 2012), suggesting they may be acquiring an iconic quality in the discursive struggle over climate change. As such, they call to mind the way images of Earth from space have come to be identified with ecological citizenship (Szerszynski & Urry, 2006). In the case of the *Jetsetter* and *Escape* entries for the Great Barrier Reef, readers were informed that it was the only living thing that could be seen from space, potentially adding to the symbolic power of the aerial images. Although the entries made no mention of coal, they referenced one element of the conflict over the proposed Adani mine, citing 'pollution from port development' as a problem and reporting that 'UNESCO spared the icon from an 'in danger' listing this year, but warns that significant conservation progress must be made' (Olander, 2015). In these entries, then, image and linguistic text harmonised to contextualise the Reef as a global asset held in trust by governments whose stewardship was monitored by an international authority on behalf of a transnational public.

Conclusion

The results of our study suggest last-chance travel lists are deeply ambivalent about climate change and environmental conflict. There is no doubt that climate change has emerged as a powerful linguistic presence in these lists, featuring in 45% of entries and more than 40% of destinations in our sample. Our data bear out the views of the journalist quoted at the start of this chapter that representing the destruction of a destination as immanent enables practitioners to acknowledge environmental problems in the traditionally celebratory genre of travel journalism. Consistent with earlier research (McGaurr, 2014, 2015), however, our study has found almost no evidence that this also creates opportunities for travel journalists to raise the material and moral dilemma of greenhouse gas emissions from long-haul flights—an action that would challenge the foundations of the global visitor economy.

On the subject of environmental conflict, our case study of the Great Barrier Reef revealed an unexpectedly strong acknowledgement of the Adani coal mine dispute. Nevertheless, all but one list in our sample failed to take these opportunities to make the connection between Australian coal, carbon emissions and degradation of the Great Barrier Reef explicit.

The limitations of last-chance travel lists identified in the study stem from their construction of destinations as both problems and products for readers, and their positioning of readers as consumers before publics, often through the use of images that represent the destination as unspoilt and desirable. This accords with a long and intimate association between spectacle-ised nature and commodification:

> The diverse environments that 'environmentalisms' want to look after, in other words, are often spectacular: they are strikingly and profoundly dramatic at the same time, quotidian and ordinary in the everyday complexity of the spectacles of nature. The relationship between these two realms or meanings is critical, with the latter (spectacular nature) providing the former (commodified spectacle) with the resources it commodifies and spectacle-ises and, conversely, the former (commodified spectacle) often polluting, and often attempting to extend the possibilities, engagements and affective resonances of the latter. (Goodman et al., 2016, pp. 678–679)

Last-chance travel lists spectacle-ise climate change. Textually, its impacts are represented as imminent and dramatic, yet as we have seen, research suggests a strong focus on catastrophic consequences tends to make people feel impotent rather than empowered (Wibeck, 2014). In our sample, dire linguistic predictions of disappearing global treasures were likely to be accompanied by beautiful images of what could yet be saved by individuals, industry and governments acting responsibly. However, most of these images had already been commodified and it seems likely their polysemous potential to position audiences as members of transnational publics was curtailed by headline calls to action entreating tourists to consume in haste. Counterbalancing this to some extent, the use of aerial images of the Great Barrier Reef and references to its visibility from space may indicate that some travel lists are positioning audiences as members of transnational publics invested in more than the immediate gratification and 'brag-ability' of consuming a natural wonder others might never have the opportunity to experience. It is also possible the unexpected prominence of environmental conflict over the Great Barrier Reef evident in our data indicates that the popularity of digitised, abbreviated texts frees travel publishers from reliance on hosted visits and an attendant sense of obligation to governments and jurisdictional tourism offices. Alternatively or simultaneously, it may be indicative of successful campaigning by the environmental movement. Further and more extensive investigation is needed to understand why only one list made connections between the mining of coal and damage to the Reef from climate change. A second study of last-chance lists published since the devastating coral bleaching event of 2016 may provide valuable additional insights.

Funding

Libby Lester received funding from the Australian Research Council as part of the Discovery Project-funded research 'Transnational Environmental Campaigns in the Australia-Asian Region' DP150103454 (2015–2018).

References

Anholt, S. (2007). *Competitive identity: The new brand management for nations, cities and regions.* Basingstoke: Palgrave Macmillan.

Anholt, S. (2010). *Places: Identity, image and reputation.* Basingstoke: Palgrave Macmillan.

Australian Conservation Foundation. (n.d.). Stop Adani's Carmichael coal mine: Protect life on our beautiful reef. Retrieved from https://www.acf.org.au/carmichael.

Australian Marine Conservation Society. (n.d.). Protecting Australia's ocean wildlife since 1965. Retrieved from https://www.marineconservation.org.au/

Barthes, R. (1977). Rhetoric of the image. In S. Heath (Ed.), *Image, music, text: Essays selected and translated by Stephen Heath* (pp. 32–51). London: Fontana.

Boykoff, M. (2007). From convergence to contention: United States mass media presentations of anthropogenic climate change Science. *Transactions of the Institute of British Geographers, 32*, 477–489.

Coglan, A. (2012). Last chance tourism on the Great Barrier Reef. In R. Lemelin, J. Dawson, & E. Stewart (Eds.), *Last chance tourism: Adapting tourism opportunities in a changing world* (pp. 133–149). London: Routledge.

Dawson, J., Lemelin, R., Stewart, E., & Taillon, J. (2015). Last chance tourism: A race to be last? In M. Hughes, D. Weaver, & C. Pforr (Eds.). *The practice of sustainable tourism: Resolving the paradox* (pp. 133–145). New York, NY: Routledge.

Destination Seeker (circa 2014–2015). 25 places you must visit before they disappear. Retrieved from http://www.destinationseeker.com/10-places-you-must-visit-before-they-disappear/.

Difrancesco, D., & Young, N. (2011). Seeing climate change: The visual construction of global warming in Canadian national print media. *Cultural Geographies, 18*(4), 517–536.

Dore, L., & Crouch, G. (2003). Promoting destinations: An exploratory study of publicity programmes used by national tourism organisations. *Journal of Vacation Marketing, 9*(2), 137–151.

Doyle, J. (2007). Picturing the clima(c)tic: Greenpeace and the representational politics of climate change communication. *Science as Culture, 16*(2), 129–150.

Fernando, A. (2013, October 14). Why listicles are here to stay. *Adweek*. Retrieved from www.adweek.com/fishbowlny/why-listicles-are-here-to-stay/264154?red=kw.

Fight for our reef. (2015). Coral bleaching emergency: Ditch Adani and rescue our reef. Retrieved from https://www.fightforourreef.org.au/bleachingemergency/

Fight for our reef. (2017). It's our reef. Help fight for it. Retrieved from https://www.fightforourreef.org.au/

Frew, E. (2008). *Climate change and doom tourism: Advertising destinations 'before they disappear'*. Paper presented at the New Zealand Tourism and Hospitality Research Conference, Hanmer Springs.

Frew, E. (2012). Advertising world heritage sites: Tour operators and last chance destinations. In R. Lemelin, J. Dawson, & E. Stewart (Eds.), *Last chance tourism: Adapting tourism opportunities in a changing world* (pp. 117–132). London: Routledge.

GetUp! (n.d.). A tax-payer funded reef-killing project. Retrieved from https://www.getup.org.au/campaigns/great-barrier-reef--3/killing-the-reef-with-taxpayers-money/a-taxpayer-funded-reef-killing-project?t=4QtnteW.

Goodman, M., Littler, J. Brockington, D., & Boykoff, M. (2016). Spectacular environmentalisms: Media, knowledge and the framing of environmental politics. *Environmental Communication, 10*(6), 677–688.

Greenpeace Australia Pacific. (2012). Save the reef. Retrieved from http://www.greenpeace.org/australia/en/what-we-do/climate/Save-the-Reef/

Greenpeace Australia Pacific. (2016). We brought the Great Barrier Reef to Canberra. Retrieved from http://www.greenpeace.org.au/blog/great-barrier-reef-canberra/

Gregory, K. (2016, December 2). Barrier Reef at risk of winding up on UNESCO 'danger list', Queensland Government says. *ABC News*. Retrieved from http://www.abc.net.au/news/2016-12-02/barrier-reef-at-risk-of-winding-up-on-unesco-danger-list/8085820

Hansen, A., & Machin, D. (2013). Researching visual environmental communication. *Environmental Communication, 7*(2), 151–168.

Hanusch, F. (2012). A profile of Australian travel journalists' views and ethical standards. *Journalism: Theory, Practice and Criticism, 13*(5), 668–686.

Hill, T. (2016, September 10). Last chance tourism is adding to the Great Barrier Reef's woes. Retrieved from http://www.takepart.com/article/2016/09/10/great-barrier-reef-last-chance-tourism-bleaching.

Independent Expert Panel. (2017, May 5). Reef 2050 Plan: Communique, 5 May 2017. Retrieved from http://www.environment.gov.au/system/files/pages/abff0d5e-b94d-4495-b79b-90dc52274f69/files/expert-panel-communique-5may2017.pdf

Köhler, M. (1998). From the national to the cosmopolitan public sphere. In D. Archibugi, D. Hed, & M. Köhler (Eds.), *Reimagining political community: Studies in cosmopolitan democracy* (pp. 231–251). Stanford, CA: Stanford University Press.

Lemelin, H., Dawson, J., Stewart, E., Maher, P., & Lueck, M. (2010). Last chance tourism: The boom, doom, and gloom of visiting vanishing destinations. *Current Issues in Tourism, 13*(5), 477–493.

Lemelin, H., Stewart, E., & Dawson, J. (2012). An introduction to last chance tourism. In R. Lemelin, J. Dawson, & E. Stewart (Eds.), *Last chance tourism: Adapting tourism opportunities in a changing world* (pp. 3–9). London: Routledge.

Lester, L. (2007) *Giving ground: Media and environmental conflict in Tasmania*. Hobart: Quintus.

Lester, L. (2016). Containing spectacle in the transnational public sphere. *Environmental Communication, 10*(6), 791–802.

Linder, S. (2006). Cashing-in on risk claims: On the for-profit inversion of signifiers for 'global warming'. *Social Semiotics, 6*(1), 103–132.

Mackellar, J., & Fenton, J. (2000). Hosting the international travel media—A review of the Australian Tourist Commission's Visiting Journalist Programme. *Journal of Vacation Marketing, 6*(3), 255–264.

Markham, A., Osipova, E., Lafrenz Samuels, K., & Caldas, A. (2016). *World Heritage and tourism in a changing climate*. Nairobi: United Nations Environment Programme, and Paris: United Nations Educational, Scientific and Cultural Organization.

McGaurr, L. (2014). Your threat or mine? Travel journalists and environmental problems. In F. Hanusch & E. Fürsich (Eds.), *Travel journalism: Exploring production, impact and culture* (pp. 231–248). Basingstoke: Palgrave Macmillan.

McGaurr, L. (2015). *Environmental communication and travel journalism*. Abingdon: Routledge.

McGaurr, L. (2016). The photography of debate and desire: Images, environment and the public sphere. *Ethical Space, 13*(2–3), 16–33.

Mountain, J. (2016, January 24). 7 stunning eco destinations to see before it's too late. *Inhabitat*. Retrieved from http://inhabitat.com/7-stunning-eco-destinations-to-see-before-its-too-late/7-eco-destinations-to-see-before.

Nisbet, M. (2009). Communicating climate change: Why frames matter for public engagement. *Environment, 51*, 12–23.

Olander, L. (2015, August 20). 11 places to see before it's too late. *Jetsetter*. Retrieved from https://www.jetsetter.com/magazine/393/places-to-see-before-its-too-late?ckey=.

O'Neill, S., Boykoff, M., Niemeyer, S., & Day, S. (2013). On the use of imagery for climate change engagement. *Global Environmental Change, 23*, 413–421.

O'Neill, S., & Nicholson-Cole, S. (2009). Fear won't do it: Promoting positive engagement with climate change through imagery and icons. *Science Communication, 30*, 355–379.

Piggott-McKellar, A., & McNamara, K. (2016). Last chance tourism and the Great Barrier Reef. *Journal of Sustainable Tourism*, 1–19. doi:10.1080/09669582.2016.1213849

Stop Adani (n.d.). Wrecks our climate. Retrieved from http://www.stopadani.com/why_stop_adani#wrecks_our_climate.

Swann, T., & Campbell, R. (2016). *Great Barrier bleached: Coral bleaching, the Great Barrier Reef and potential impacts on tourism* (Discussion paper). Canberra: Australia Institute. Retrieved from http://www.tai.org.au/sites/defualt/files/Swann%20Campbell%202016%20Great%20Barrier%20Bleached%20FINAL%20w%20cover.pdf.

Szerszynski, B., & Urry, J. (2006). Visuality, mobility and the cosmopolitan: Inhabiting the world from afar. *The British Journal of Sociology, 57*(1), 113–131.

Tara. (2015). 10 amazing places to see before it's too late. *Lush Palm*. Retrieved from http://lushpalm.com/10-places-to-see-before-its-too-late/

Tarte, D., Hart, B., Hughes, T., & Hussey, K. (2017). *Reef 2050 long term sustainability plan: Progress on implementation: Review by Great Barrier Reef Independent Review Group*. Retrieved from http://apo.org.au/files/Resource/pub-reef-2050-long-term-sustainability-plan-21feb17.pdf

Taylor, L. (2015, March 6). Great Barrier Reef lobbying: Australian government offers junkets to journalists. *The Guardian* (Australian Ed.). Retrieved from https://www.theguardian.com/environment/2015/mar/06/great-barrier-reef-lobbying-australian-government-offers-junkets-to-journalists.

UNESCO. (1972). *Convention concerning the protection of the world cultural and natural heritage*. Retrieved from http://whc.unesco.org/en/conventiontext/.

Wibeck, V. (2014) Enhancing learning, communication and public engagement about climate change—Some lessons from recent literature. *Environmental Education Research, 20*(3), 387–411.

Yaccato, B. (circa 2015). 20 places to see before it's too late. *EscapeHere*. Retrieved from http://www.escapehere.com/inspiration/20-places-to-see-before-its-too-late/

9. The Representation of Risk in the British Press: Comparing Reporting of Climate Change and Terrorism

Victoria Dando

Introduction

This chapter aims to understand media coverage of climate change by situating it in the wider context of the reporting of global risks, comparing the coverage of climate change with coverage of terrorism. Drawing on a longitudinal analysis of more than 1,000 newspaper articles between 1999 and 2012, the chapter examines the discursive and ideological conditions involved in the construction of climate change and terrorism in five British newspapers. Focusing on the representation of these issues as risks, it argues that the media have normalised the terrorist threat as a current and real issue for contemporary British society, overemphasising the risk facing the UK and the West while largely ignoring other areas of the world. In contrast, climate change is presented as a future risk. While the West features in reports, it is without the intensity and focus of that found in terrorism news, disconnecting the issue from a dominant national discourse.

Media Realised Risks

According to Ulrich Beck's 'World Risk Society', modern-day society is now in the grip of global risks produced by the success of the past (Beck, 2001). Such risks include climate change and terrorism—the former a product of successful industrialisation (Beck, 2009b, p. 8) and the latter which 'annuls the rational principles of former risk calculations' (Beck, 2009b, p. 14). While risk society is not necessarily more hazardous (Beck, 2001), the nature of risk has changed. External threats were once considered as characterised by fate or God (Beck, 2009, p. 4), whereas risks are now manmade, unpredictable, and invisible with consequences for society globally.

At the heart of World Risk Society is the mass media, a central conduit in the '"experienceability" and "reality" of global risks' (Beck, 2009, p. 72). For the vast majority of people, the very 'reality' of climate change and terrorism is dependent on how they are articulated, defined, and debated in the media (Beck, 2009); thus, the way the media tell the story of these risks matters (Lewis, 2012, p. 268).

To date, and when measured against the certainty of the science and the compelling need for action, there is considerable disparity in media coverage of climate change. Climate change has tracked along an inconsistent news trajectory, fitting awkwardly with news agendas (Lewis, 2012, p. 260). The contours of that coverage have been shaped largely by politics (Boykoff, 2008; Boykoff & Boykoff, 2004; Carvalho & Burgess, 2005), key climate events such as the 1997 Kyoto Protocol (Boykoff & Roberts, 2007) and 'weather events [and] charismatic megafauna' (Boykoff, 2008, p. 549). Such news hooks fail to reflect the reality: climate change is a sweeping policy issue affecting not only the environment, but with the potential to ignite changes in social and economic stability, public health, and spark conflict and mass migration in the world's poorest and most volatile regions (see Stern, 2006, and each of the Assessment Reports from the Intergovernmental Panel on Climate Change (IPCC)).

Despite its comprehensive evidence base, alongside recommendations that only a 'green revolution' will suffice in responding to the magnitude of the risk (Green & Stern, 2014), climate change has failed to resonate with news agendas (Lewis & Boyce, 2009; Smith, 2005). Instead, in the post 9/11 world, it is terrorism that has calibrated political and public perspective on what constitutes risk. This is despite data indicating that terrorist attacks in the West are less frequent than at previous times in history (see for example the Global Terrorism Database). Studies also indicate that international terrorism is a minor threat when compared to other serious global risks (Abbott et al., 2006, p. 4). Nonetheless, terrorism has received significant and sustained media coverage, commanding more attention, more urgency and less scrutiny than climate change (Lewis, 2012, p. 260). Terrorism has become 'the most serious threat in our collective imagination' (Freedman & Thussu, 2012, p. 2).

In order to understand more about these disparities, this study explores the discursive and ideological construction of climate change and terrorism as risks. It departs from previous research in the field of journalism, media, and communication by comparing both issues simultaneously and through the same framework. It examines the dominant media narratives around risk and effect, teasing out what they reveal, conceal, and distort (Lewis, 2012, p. 268), and explores what those advocating action on climate change can learn from the terrorism media narrative.

Method

This chapter presents the findings from a longitudinal, comparative content analysis of reporting of climate change and terrorism in the UK, looking at the *Daily Mail*, *The Guardian*, *The Independent*, *Daily Mirror*, and *The Times.* These newspapers represent a cross-section of the British newspaper industry, spanning the political spectrum and the tabloid-broadsheet divide. The research covers a 14-year period from January 01, 1999 to December 31, 2012, comprised of two separate data sets: the first examining articles from January 1999 to December 2008, which was the initial scope of the work; and a second, which examined articles from January 2009 to December 2012 to add timeliness. To gather the sample, a search of *LexisNexis* was carried out for climate change and terrorism separately, using the keywords 'climate change' or 'global warming'; and 'terrorism', 'terrorist' or 'terrorists'. Articles with three or more mentions of any of these keywords was set as a search term.

This generated 12,162 articles for climate change and 35,126 articles for terrorism—an early indicator that of the two issues, the media attach greater seriousness and significance to terrorism. From this a smaller sample was gathered for further analysis, giving 496 articles for climate change and 517 for terrorism.

In the field of journalism, content analysis has been used to examine the impact of reporting of terrorism on policy (Weimann & Brosius, 1991); longitudinal studies of climate change (Boykoff, 2007; Carvalho & Burgess, 2005; McComas & Shanahan, 1999); the portrayal of Muslims in the British press (Moore et al., 2008); and to identify the ways in which media coverage undermined or validated the British government's case for war in Iraq (Lewis & Brookes, 2004), among others.

Importantly, content analysis can cope with large volumes of data (Krippendorff, 2004; Riffe et al., 2005; Stemler, 2001) and caters for longitudinal studies, both of which were key criteria in this research. The method enables trends and patterns to emerge and the evolution of media coverage to be documented (Hansen et al., 1998; Krippendorff, 2004; Stemler, 2001).

Articles were coded for two areas: risk and effects. The aim was to gather data on the urgency of risk and effects in news narratives. Risk was coded based on timeframe—current risk, future (anticipated) risk, or both. Data on the timeframe of effects—past, present, or future (along with a category of 'both' which referred to present and future)—was collected, along with whether those effects were reported as positive, negative, both or debated. The location affected was also recorded.

Reporting Risk: 'Normalising' Western Terror Attacks and Distancing Major Climate Risk

77% of articles reporting risk referred to climate change as an anticipated or future risk, 13% as a current risk, and 10% as both a current and future risk. By contrast, 52% of articles reporting terrorism risk focus on current risk, 41% as an anticipated or future risk, and 7% as both.

What these data show is that significant climate risk is situated in the future in news reports, distanced from the present day. While the IPCC has documented future climate change extensively—and as such news coverage of this timeframe is not incorrect—its reports also evidence modern day risks such as extreme weather events and shrinking glaciers (see, for example, the Fifth Assessment Report). The current impact of climate change is also outlined in reports detailing deaths from climate change impacts (400,000 people a year according to DARA's 2012 *Climate Vulnerability Monitor*). When it was reported, current risk included references to droughts in Africa (*The Guardian*, October 24, 2005), the melting of the ice in the Arctic (*The Independent*, November 21, 2000), and the impact of changing temperatures on the occurrence of powerful natural disasters (*The Times*, September 14, 2004). Declarations of 'unprecedented change' (*Daily Mirror* (feature), April 12, 2007) and a 'planetary emergency' (*Greenpeace quote, Daily Mail*, September 22, 2007) were also found in reports, *but did not translate into a sustained current risk narrative.* It is striking too, that the potentially massive refugee crises likely to be caused by climate change is largely absent from news reports in this study. Although figures are mentioned occasionally in headlines—'50m environmental refugees by end of decade, UN warns' (*The Guardian*, October 12, 2005)—the scale of this crisis is at odds with its prominence.

In contrast, there is a powerful narrative that emphasises—and potentially over-inflates—the risk of terrorism as a current, modern day phenomenon. Alongside actual terrorist attacks, reports referenced the threat of imminent attacks, creating a perpetual and immediate sense of risk. There was a 10-fold increase in the number of articles referencing the current risk of terrorism between 1999 and 2001, the latter being the year of terrorist attacks on the World Trade Centre and Pentagon in America. There are rises and falls in the following years, with peak occurrences of current risk recorded in 2004, the year of the Madrid train bombing, 2005, the year of the 7/7 attacks in London, and 2008, following attacks in Mumbai. Articles reported flight suspensions following 'credible intelligence of a serious threat to UK aviation interests in Saudi Arabia' (*The Times*, August 14, 2003) and quotes from then Prime Minister Tony Blair saying 'there was no longer any "leafy suburb" safe from weapons of mass destruction, religious fanaticism and terrorism' (*Daily Mirror*, May 27, 2002).

While this current risk narrative is able to steer news agendas, it is at odds with official data on terrorist attacks in the West. According to the most recent Global Terrorism Index (2016), Iraq is the country most likely to experience terrorism; the United Kingdom and America rank 34th and 36th in the table respectively. The Index states that in 2015, 29,376 people died around the world from terrorist attacks, considerably less than the figure for climate change, but this disparity is not reflected in either the quantity of coverage or the focus on current risk.

When we look at how each of the newspapers reported risk, a pattern emerges that indicates the ability of terrorism to act as an ideological unifier. With the exception of the *Daily Mirror*—which overwhelmingly focused on the present-day—each of the newspapers in this study reported current and future terrorism risk in similar frequencies (see Figure 9.1). In contrast, climate risk reporting is generally confined to the broadsheet rather than the tabloid press (see Figure 9.2). Ideology also appears to have an impact among the broadsheet newspapers' coverage of climate change.

The Guardian included almost double the amount of coverage of climate change as an anticipated risk compared to *The Independent* and *The Times*.

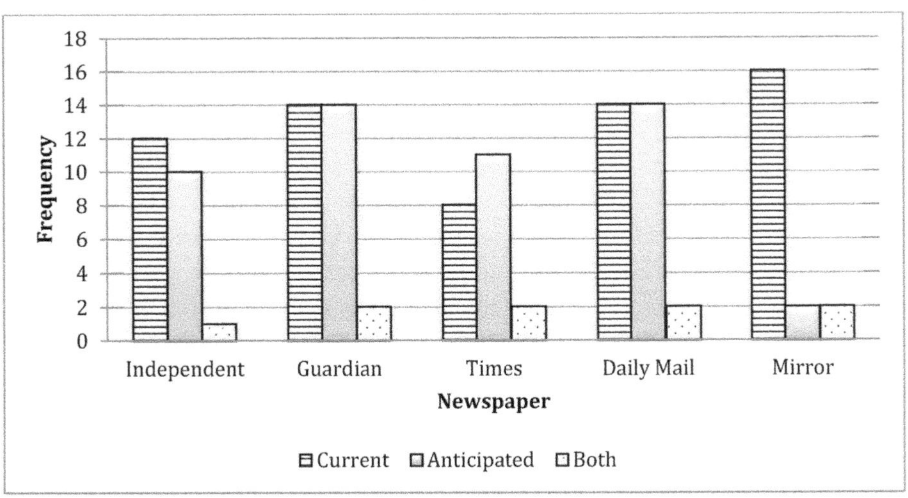

Figure 9.1: Newspaper coverage of terrorism risk (1999–2012).

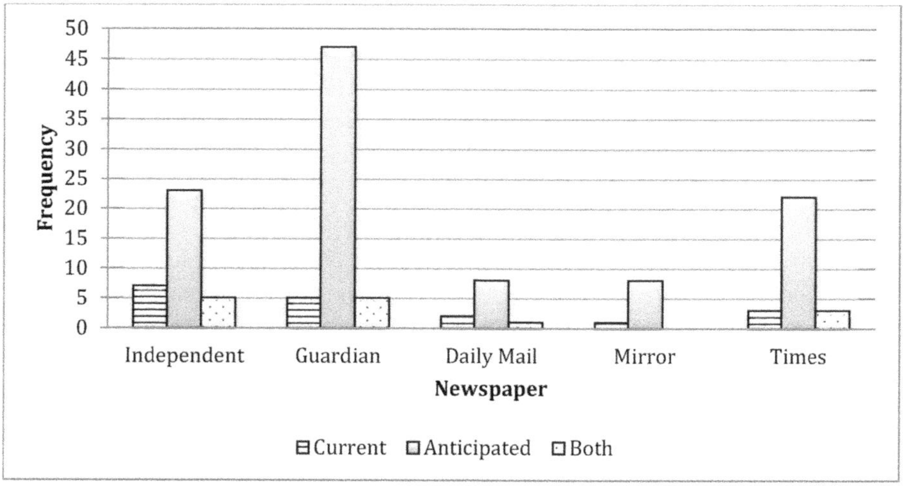

Figure 9.2: Newspaper coverage of climate change risk (1999–2012).

Conservative politics and scepticism of climate change are strongly linked (Antonio & Brulle, 2011; Corner, 2013; Jacques et al., 2008; McCright & Dunlap, 2010, 2011; Poortinga et al., 2011), and the lack of focus on risk in *The Times*—and to an extent *The Independent*, which has no party political line (Carvalho & Burgess, 2005, p. 1460)—*indicates* that this attitude towards climate change may influence the visibility of stories on the subject. Here, there are implications for balanced, comprehensive and representative coverage that promotes informed public debate.

The data in Figure 9.2 point towards a normalisation of *major* terrorist risk in British news narratives as a current issue, in spite of evidence to the contrary. This suggests that when risk is reported, there is an over-reliance on events such as 9/11 and 7/7 to not only perpetuate coverage, but to position terrorism as an ongoing threat. The scientific consensus on climate risk appears to be more selectively reported. While future climate risk is an issue, there are opportunities to highlight the current significant problems facing those at the coalface of climate change in the Global South—and other areas around the world—but this was found to be less of an angle for the newspapers in this study.

Reporting Effects: Focusing on the West and Ignoring the Rest—the Difference Between Climate Change and Terrorism News

Minor differences were found in the timeframes used to report the *effects* of climate change. Of the articles referencing effects, 35% referred to future effects, 34% to current effects, 28% to both, and 3% to past effects. The effects of terrorism are reported predominantly in the past, with 64% of articles referencing effects referring to this timeframe and 34% of articles referring to the current effects of terrorism.

What this data shows is that while newspaper narratives represent the disparate nature and timeframe of climate change effects, when the issue is compared to terrorism, little attention is paid to past effects, for example severe flooding in the UK in 2000 and 2007, the world's first 'climate conflict' in Darfur (according to the UN Environment Programme report in 2007), and hurricane Katrina in New Orleans in 2005. They don't however, retain continuity of coverage in the same way that 9/11, 7/7, and other attacks on the West have done. While these terrorist attacks have become reference points for reporting and society more widely (Mythen & Walklate, 2006), the same isn't true for significant climate change events, which rarely transcend the 'now'. While terrorism news uses, for example, anniversaries of attacks, new terrorist threats and political action to repurpose details of effects, these elements of the immediate modern past are largely missing from the long-term climate change narrative.

There is a clear signal from all five newspapers of the longevity of terrorist attacks as news, and a consistent use of the past (see Figure 9.3). *The Times* used the past most often of all newspapers in its reports, with *The Guardian* and *The Independent* reporting on both past and current effects in similar numbers. When it comes to climate change, *The Guardian* and the *Daily Mirror* were the only two newspapers to situate the effects in the future more

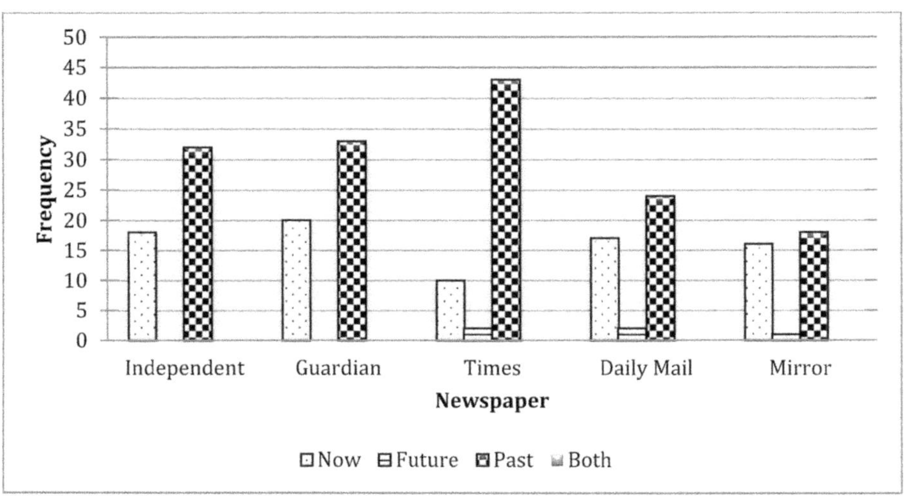

Figure 9.3: Effects of terrorism as reported by newspaper (1999–2012).

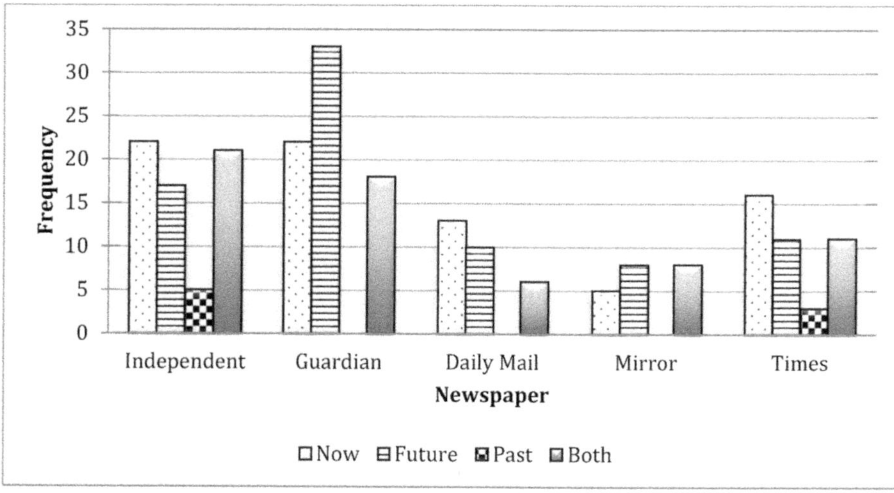

Figure 9.4: Effects of climate change as reported by newspaper (1999–2012).

than the present (or past), however the *Daily Mirror's* coverage of effects is scarce overall (see Figure 9.4). *The Times*, *The Independent* and the *Daily Mail* focused on current effects most often. Among the three broadsheet newspapers, there are fewer articles in *The Times* that address the effects of climate change overall. As with risk, climate effects aren't a strong news agenda for the right-leaning *Times*.

As we might expect, the effects of both climate change and terrorism are generally reported as negative. We find much more equivocation, however, in the reporting of climate change. 98% of articles about terrorism refer to negative effects, compared with 79% of articles which outlined the negative impact of climate change (such as flooding, species loss and environmental refugees). A substantial minority of articles take a more sanguine view of climate change: 4% reported positive effects (such as warmer temperatures, fewer days in hospital because of better health), 4% detailed both within articles, and 13% of articles debated about whether effects would be positive or negative.

This contrasts with the view of most scientists, for whom the warming of the climate system is unequivocally negative (IPCC, 2013, p. 4). At its unchecked worst, climate change has the capacity to 'alter the sustainability of life on earth as we know it' (Lewis, 2012, p. 260), transforming the physical geography of the world with implications for where and how people live (Stern, 2006, p. vi). This data shows that reporting of climate change lacks the same level of intensity as terrorism in this area and there is more debate about the outcome of the effects.

Telling differences emerge too, when we look at the tabloid's treatment of climate change when compared to terrorism. The *Daily Mail* and *Daily Mirror* were much less likely to include details of the negative effects of climate change in their reports, with 9% and 8% respectively including such details. These figures are compared to 19%, 28%, and 36% respectively for *The Times, The Independent* and *The Guardian. The Times'* result here, is also suggestive of the impact of conservative ideology on its reporting of the impact of climate change effects. However, when it came to detailing the negative effects of terrorism, the *Daily Mail* and *Daily Mirror* reporting is much closer to the broadsheets—15% (*Mirror*), 18% (*Daily Mail*), 22% (*Independent*), 22% (*Guardian*) and 23% (*Times*)—with very little evidence of any ideological inflection.

So while we see a uniformity of coverage on the negative effects of terrorism, on climate change we see disparities between the popular and the prestige press, and between the right-leaning and left-leaning broadsheets, where there appears to be a division in the visibility of climate change as a negative issue. The problem here, as Boykoff & Mansfield (2008, p. 7) point out, is that the readers of tabloid newspapers are important in terms of pressure to move climate policy forward. If the effects are discussed less in these newspapers, there are also fewer opportunities to focus on the negative impact of the issue, which has clear implications for the success of future action (Boykoff & Mansfield, 2008). The same is true for right-leaning newspaper like *The Times.* If negative impacts are marginalised or ignored, the significance of the issue is not only undermined, but avenues to engage the centre-right in discussions about action are closed off.

A further pattern emerges from data gathered about where in the world the effects of terrorism and climate change are felt. When effects are reported, 31% of articles referenced Western nations as experiencing the effects of climate change either in the past, present, or future; 25% referred to the developing world/Global South; 27% to the world as a whole while 11% of articles focused on the impact on wildlife and 6% on the Polar Regions. There is a clear spread of affected areas representative of the global impact of climate change. Here, though, the tone and content of those articles is varied.

In relation to the West, there were examples of reports accentuating the unusual or unexpected in relation to climate change effects—a narrative largely absent in reports about the developing world/Global South. While these are negative, such as the loss of habitats or the appearance of flowers earlier, the focus on flora and fauna (among other areas) contribute to a narrative of 'ordinary' effects of climate change, which are also reported at times with humour. This focus dissipates the significant risks facing the West, which are well documented by the IPCC (e.g. see 2014a, 2014b). Hamblyn (2009, p. 235) refers to a 'rhetorical mismatch' between apocalyptic climate warnings and 'cheerful' policy actions; but this research shows such a mismatch is also evident in the reporting of climate effects (and risk) for the West.

Droughts in Africa and the warming of the environment in the Himalayas were some of the more serious current, negative effects of climate change reported. The melting of the ice caps in the Arctic and Antarctic also featured. Research by Farbotko and Lazrus (2012) showed how residents of low-lying islands such as Tuvalu in the Pacific, have come to symbolise climate change, held up by Western politicians and the media as the first 'victims' or 'refugees' of climate change. While this study reinforces that focus, it also finds some evidence of hard-hitting references to the effects of climate change in Britain and the West. Flooding (and the threat of it), particularly in relation to severe flooding in the UK in 2000 and 2007, featured in reports. 'The rain still pours down like Noah's flood ... Seven people are dead and 27,000 homes have been flooded, along with 5000 businesses. Maybe 100,000 people in all will return to stinking, filthy homes where the electricity and water doesn't work, the drains have backed up, the stink has seeped into the plaster which needs ripping out and floors buckle,' reported *The Guardian* in a comment piece about the floods in Britain in 2007 (July 3, 2007). In an editorial *The Independent* stated: '... most of Australia is already suffering from a decade-long drought ... agriculture has been devastated ... even Perth on the west coast, are struggling to maintain supplies' (July 25, 2007).

While terrorism is (like climate change) much more likely to affect people outside the West, its reporting is much more likely to focus on effects closer to home. Western people and places were reported as witnessing and experiencing the effects of terrorism in 80% of references, compared to 17% for Middle Eastern places or people, and the 'world' 3% of the time. The developing world did not feature at all. This data shows that *there is a clear Western narrative around the impact and effects of terrorism* that includes topics such as enhanced security and military effects, psychological trauma and community or national disruption. Climate change and terrorism share common ground when it comes to those areas of the world affected by each issue. So while climate change reporting does, to some extent, reflect its global character, the coverage of terrorism is much more likely to report it as a Western problem.

Discussion

Beck says that without the media, risks are nothing at all (2006, p. 332). Clearly, some risks are not as newsworthy as others (Lewis, 2012, p. 260) but the claim that climate change is the 'most severe problem that we are facing today-more serious even than the threat of terrorism,' (King, 2004) is not reflected in media coverage. This research has showed how five British newspapers have helped to generate, endorse, and sustain the risk of terrorism while creating a much more equivocal narrative about the risk of climate change.

Terrorism has been uniformly constructed in the British press as a risk that is current and close at hand. Terrorist attacks in the West since 9/11 have become enduring news hooks and a discursive shorthand for risk. The use of history in the media representation of terrorism has been shown to have an effect on bridging divides and resonating with the public (Winfield et al., 2002). At the same time the dominance of the modern past in news reports about terrorism in this study may have served to increase the relevance of the issue and its topicality—a narrative that both distorts and conceals certain realities. Acts of terrorism and terrorism-related violence occur much more frequently in the developing world, but were barely reported—instead terrorisms risks and effects are much more concentrated and characterised as a British and Western issue. Together, this has discursively condensed the terrorist threat into a single, Islamic-based entity (Freedman & Thussu, 2012; Jayyusi, 2012), removing complexities around international terrorism. It is also clear that the consistent and similar portrayal of terrorism (in terms of visibility) as a risk in all of the newspapers in this study, along with a cohesive discourse around the negative effects of the issue, signal its significance to audiences. Compared to the real risks that people face, this presents something of an imbalance since '[m]ost of us live at peace in democracies' (Briscoe, 2008, p. 81). There has, in this sense, been an over-inflation of the terrorist threat (Lewis, 2012, p. 258) amidst the continuity of an ever breaking story (Freedman & Thussu, 2012, p. 2; Zalman & Clarke, 2009).

In comparison to terrorism, climate change risk lacks a distinct British—or Western—hallmark. While climate risks and effects facing the West are reported, this lacks the same focus and commitment as terrorism news. British newspapers sustain narratives that associate dangerous climate change primarily with other people, places, and species. Dangerous climate change is also distanced in narratives, reported primarily as an event for the future; people must therefore think of a world beyond their own lives and envision an intergenerational link with family and friends they may never meet or know. Previous studies have found that climate change risks are socially defined as issues primarily for developing countries, indigenous people and ways of life, and certain species of wildlife (e.g. Boykoff, 2008; Doulton & Brown, 2009; Farbotko & Lazrus, 2012, p. 4; Lorenzoni et al., 2007). While these personify climate change, they also serve to demarcate the boundaries of the issue, removing it from the UK. In this study, while the UK and West are not absent from reports about climate change, they do not have the same dominant, recurrent—and limited—British narrative as those in terrorism reports. The end result is that, when compared to terrorism, climate change is 'undone' by the lack of an overarching strong, nationally-relevant risk and effect narrative in the British press.

What this research has also shown is that while the risk of terrorism is reported universally, the risk of climate change is reported through an ideological lens. While the coverage of terrorism may act as an ideological trope (Freedman & Thussu, 2012, p. 12), it traverses political divides across newspaper titles. The result is a sustained—somewhat distorted—relevance that drives news coverage. In contrast, when *The Guardian* and *The Times'* coverage of climate change are considered together, the impact of ideology is particularly apparent. There is a noticeable split in the frequency with which the right-leaning *Times* discusses the effects of climate change compared to *The Guardian* (and *The Independent)*. We know from existing research that weather, flora and fauna, as tools to convey the current and future impact of climate change to audiences, may alienate centre-right individuals (Corner, 2013). We also know that there are links between conservative ideology and climate change scepticism, doubt, and lack of trust (see Antilla, 2005; Leiserowitz et al., 2014; McCright & Dunlap, 2003). This research further adds to that data by illustrating how climate change is less likely to engage the centre-right. Political divides on the issue affect news narratives in the centre-right marginalising coverage of both risk and effects.

It is also clear that there is a prestige-popular press divide on climate change that is not present with terrorism. Tabloids are less likely to report on the negative impact of climate change, meaning that news fails to capture the enormity and urgency of the issue. Climate change permeates every aspect of life (Heal & Kriström, 2002, p. 6), but this research shows that less focus on risk and effects in the tabloids potentially constructs a restricted view of climate change.

In summary, the data here suggests that narratives are failing to communicate the scale, relevance, and urgency of climate change to British audiences, while amplifying those messages in relation to terrorism. The issue is more apparent in the tabloid and centre-right media coverage of climate change. Dominant discursive timeframes that convey immediacy and relevance are, in contrast naturalised in relation to terrorism, across all newspapers. Such rhetoric reflects the current position on political reaction to each issue, as Lewis notes: there has been 'action in one instance and inertia in another'

(Lewis, 2012, p. 262), responses 'entirely in keeping with the urgency of the news coverage' (2012, p. 262) on climate change and terrorism, and demonstrating how the news is not 'politically innocent' (2012, p. 260). Cottle too (2009, p. x) notes the 'fundamental disconnect' in media coverage of climate change and the political response needed.

The comparative approach used in this research brings discussions around climate change communications into sharper relief. It situates climate change in the context of wider major global risks, and illustrates the differences in terrorism and climate change news narratives. The research suggests potential avenues for climate change communicators to consider in communicating climate change risks. Corner suggests that 'climate change must break out of its left-wing ghetto in order for a new, meaningful conversation with the centre-right to begin' (2013, p. 2). The data here lends weight to calls for a climate narrative that engages the centre-right, but also one that can negotiate media boundaries. Groups such as Climate Outreach—Europe's leading climate change communicators—point to catastrophe narratives and doom and gloom as one of the downfalls in climate communication (Corner, 2017). Existing research has also suggested fearful and threatening representations of climate change are inappropriate drivers for engagement (O'Neill & Nicholson-Cole, 2009). Yet, terrorism seems to have thrived in the media on such a discourse. Indeed on engagement, terrorism is named by members of the public in opinion polls as one of the most important issues facing the UK, well ahead of the environment (see, for example Ipsos MORI February 2017 Issues Index). There are calls for more use of 'alarming' lexical choices and frames in climate communication. Good (2008, p. 247) suggests that greater benefit may stem from terms that elaborate on the catastrophe element of climate change, such as 'climate catastrophe' or 'climate cancer'. Corbett & Durfee (2004, p. 144) also venture into this territory, arguing that the phrase 'global warming' lacks seriousness. Based on the cohesiveness of the terrorism narrative in communicating catastrophic risk, this research suggests that such an avenue may be of value in communicating the seriousness of climate change.

A combination of use of the past along with a real focus on Britain and the West in the terrorism narrative has given the issue a distinct 'home' and largely succeeded in overcoming ideological and popular/quality differences. Adopting such a focus, along with a strong representations of climate risk could significantly benefit the telling of the climate change story. In fact, Climate Outreach has already suggested in its publication *Managing the Psychological Distance of Climate Change*, that climate change is distant, not 'here' and not 'now' (2016, p. 1). They outline a number of methods to boost the relevance of climate change, including the use of past and previous experiences to reduce the psychological distance with the issue—although they note that caution is needed to ensure efforts don't backfire (2016, p. 2). Climate change, while an issue for the world, is shown to be something of a narrative nomad by this research. If climate change is to learn from terrorism coverage, more can be done by climate communicators—and the media—to bring climate change closer to home (2016, p. 1). Making it nationally relevant, highlighting tangible impacts, and actually emphasising risk may help build and sustain a compelling news rhetoric that more accurately reflects the scientific consensus and engage audiences on the issue.

References

Abbott, C., Rogers, P. & Sloboda, J. et al. (2006). *Global responses to global threats. Sustainable security for the 21st century.* Oxford: Oxford Research Group.

Antilla, L. (2005). Climate of scepticism: US newspaper coverage of the science of climate change. *Global Environmental Change-Human and Policy Dimensions, 15*(4), 338–352.

Antonio, R. J., & Brulle, R. J. (2011). The unbearable lightness of politics: Climate change denial and political polarization. *The Sociological Quarterly, 52*, 195–202.

Beck, U. (2001). *World risk society.* Cambridge: Polity Press.

Beck, U. (2006). Living in the world risk society. *Economy and Society, 35*(3), 329–345.

Beck, U. (2009). *World at risk.* Cambridge: Polity Press.

Boykoff, M. T. (2007). Flogging a dead norm? Newspaper coverage of anthropogenic climate change in the United States and United Kingdom from 2003 to 2006. *Area, 39*(4), 470–481.

Boykoff, M. T. (2008). The cultural politics of climate change discourse in UK tabloids. *Political Geography, 27*(5), 549–569.

Boykoff, M. T., & Boykoff, J. M. (2004). Balance as bias: Global warming and the US prestige press. *Global Environmental Change-Human and Policy Dimensions, 14*(2), 125–136.

Boykoff, M. T., & Mansfield, M. (2008). 'Ye Olde Hot Aire': Reporting on human contributions to climate change in the UK tabloid press. *Environmental Research Letters, 3*(2), 1–8.

Boykoff, M. T., & Roberts, J. T. (2007). *Media coverage of climate change: Current trends, strengths, weaknesses.* Human Development Report 2007/2008, Occasional paper.

Briscoe, S. (2008). Panic now! Or later? *Significance, 5*(2), 79–81.

Carvalho, A., & Burgess, J. (2005). Cultural circuits of climate change in UK broadsheet newspapers, 1985–2003. *Risk Analysis, 25*(6), 1457–1469.

Climate Outreach. (2016). *Managing the psychological distance of climate change.* Retrieved from http://climateoutreach.org/resources/psychological-distance/

Corbett, J. B., & Durfee, J. L. (2004). Testing public (un)certainty of science: Media representations of global warming. *Science Communication, 26*, 129–151.
Corner, A. (2013). *A new conversation with the centre-right about climate change: Values, frames and narratives.* Oxford: Climate Outreach and Information Network.
Corner, A. (2017, March 10). Blog: Prince Charles' Ladybird book is a climate communications masterpiece. *Climate Outreach.* Retrieved from http://climateoutreach.org/blog-prince-charles-ladybird-book-climate-communications-masterpiece/
Cottle, S. (2009). Series editors preface. Global crises and the media. In T. Boyce & J. Lewi (Eds.), *Climate change and the media* (pp. vii–xi). New York, NY: Peter Lang.
DARA and the Climate Vulnerable Forum. (2012). *Climate vulnerability monitor 2nd edition. A guide to the cold calculus of a hot planet.* Retrieved from http://daraint.org/wp-content/uploads/2012/10/CVM2-Low.pdf
Doulton, H., & Brown, K. (2009). Ten years to prevent catastrophe? Discourses of climate change and international development in the UK press. *Global Environmental Change—Human and Policy Dimensions, 19*(2), 191–202.
Farbotko, C., & Lazrus, H. (2012). The first climate refugees? Contesting global narratives of climate change on Tuvalu. *Global Environmental Change, 22*(2), 382–390.
Freedman, D., & Thussu, D. K. (2012). Dynamics of media and terrorism. In D. Freedman & D. K. Thussu (Eds.), *Media and terrorism: Global perspectives* (pp. 1–20). London: Sage.
Global Terrorism Index. (2016). Retrieved from http://economicsandpeace.org/wp-content/uploads/2016/11/Global-Terrorism-Index-2016.2.pdf
Good, J. E. (2008). The framing of climate change in Canadian, American, and international newspapers: A media propaganda model analysis. *Canadian Journal of Communication, 33*, 233–255.
Green, F., & Stern, N. (2014). *An innovative and sustainable growth path for China: A critical decade.* Policy paper. Centre for Climate Change Economics and Policy Grantham Research Institute on Climate Change and the Environment.
Hamblyn, R. (2009). The whistleblower and the canary: Rhetorical constructions of climate change. *Journal of Historical Geography, 35*, 223–236.
Hansen, A. et al. (1998). *Mass communication research methods.* Basingstoke: Macmillan.
Heal, G., & Kriström, B. (2002). Uncertainty and climate change. *Environmental & Resource Economics, 22*(1–2), 3–39.
IPCC. (2013). *The physical science basis.* Summary for policymakers.
IPCC. (2014a). *Impacts, adaptation, and vulnerability.* Summary for Policymakers.
IPCC. (2014b). Climate change 2014: Synthesis report. Contribution of Working Groups I, II and III to the Fifth Assessment Report of the Intergovernmental Panel on climate change [Core Writing Team, R. K. Pachauri & L. A. Meyer (Eds.)]
Ipsos MORI. (2017). *February issues index.* Retrieved from https://www.ipsos-mori.com/researchpublications/researcharchive/3838/Concern-about-the-NHS-jumps-to-the-highest-level-since-2003.aspx

Jacques, P. J. et al. (2008). The organisation of denial: Conservative think tanks and environmental scepticism. *Environmental Politics, 17*(3), 349–385.

Jayyusi, L. (2012). Terror, war and disjunctures in the global order. In D. Freedman & D. K. Thussu (Eds.), *Media and terrorism: Global perspectives* (pp. 23–46). London: Sage.

King, D. (2004, January 9). Global warming 'biggest threat'. *BBC News online*. Retrieved from http://news.bbc.co.uk/1/hi/sci/tech/3381425.stm

Krippendorff, K. (2004). *Content analysis: An introduction to its methodology*. London: Sage.

Leiserowitz, A., Maibach, E., Roser-Renouf, C., Feinberg, G., & Rosenthal, S. (2014) *Politics & Global Warming*. Yale University and George Mason University. New Haven, CT: Yale Project on Climate Change Communication.

Lewis, J. (2012). Terrorism and news narratives. In D. Freedman & D. K. Thussu (Eds.), *Media & terrorism: Global perspectives* (pp. 257–270). London: Sage.

Lewis, J., & Boyce, T. (2009). Climate change and the media: The scale of the challenge. In T. Boyce & J. Lewis (Eds.), *Climate change and the media* (pp. 3–16). New York, NY: Peter Lang.

Lewis, J., & Brookes, R. (2004). How British television news represented the case for the war in Iraq. In S. Allan & B. Zelizer (Eds.), *Reporting war: Journalism in wartime* (pp. 283–300). Oxford: Routledge.

Lorenzoni, I. et al. (2007). Barriers perceived to engaging with climate change among the UK public and their policy implications. *Global Environmental Change-Human and Policy Dimensions, 17*(3–4), 445–459.

McComas, K., & Shanahan, J. (1999). Telling stories about global climate change—Measuring the impact of narratives on issue cycles. *Communication Research, 26*(1), 30–57.

McCright, A. M., & Dunlap, R. E. (2003). Defeating Kyoto: The conservative movement's impact on U.S. climate change policy. *Social Problems, 50*(3), 348–373.

McCright, A. M., & Dunlap, R. E. (2010). Anti-reflexivity. The American conservative movement's success in undermining climate science and policy. *Theory, Culture & Society, 27*(2–3), 100–133.

McCright, A. M., & Dunlap, R. E. (2011). Cool dudes: The denial of climate change among conservative white males in the United States. *Global Environmental Change,* 21 (4), 1163–1172

Moore, K. et al. (2008). *Images of Islam in the UK: The representation of British Muslims in the national print news media 2000–2008* (Working Paper). Cardiff: Cardiff University.

Mythen, G., & Walklate, S. (2006). Communicating the terrorist risk? Harnessing a culture of fear? *Crime, Media, Culture, 2*(2), 123–142.

O'Neill, S., & Nicholson-Cole, S. (2009). Fear won't do it. Promoting positive engagement with climate change through visual and iconic representations. *Science Communication, 30*(3), 355–379.

Poortinga, W. et al. (2011). Uncertain climate: An investigation into public scepticism about anthropogenic climate change. *Global Environmental Change, 21*(3), 1015–1024.

Riffe, D. et al. (2005). *Analysing media messages.* Mahwah, NJ: Lawrence Erlbaum Associates.

Smith, J. (2005). Dangerous news: Media decision making about climate change risk. *Risk Analysis, 25*(6), 1471–1482.

Stemler, S. (2001). An overview of content analysis. *Practical Assessment, Research & Evaluation, 7*(17) [online]. Retrieved from http://pareonline.net/getvn.asp?v=7&n=17

Stern, N. (2006). *The economics of climate change.* Cambridge: Cambridge University Press.

Weimann, G., & Brosius, H. B. (1991). The newsworthiness of international terrorism. *Communication Research, 18*(3), 333–354.

Winfield, B. H. et al. (2002). History as the metaphor through which the current world is viewed: British and American newspapers' uses of history following the 11 September 2001 terrorist attacks. *Journalism Studies, 3*(2), 289–300.

Zalman, A. & Clarke, J. (2009). The global war on terror: A narrative in need of a rewrite. *Ethics & International Affairs, 23,* 101–113.

10. *The Media Produce Climate Change: Fake News, Post-Truth and Journalistic Pollution*

RICHARD MAXWELL AND TOBY MILLER

The Media Produce Climate Change!

That's the kind of title we'd have expected to see at the high point of post-modern approaches to studying the media—stirring and even shocking intellectual innovations of the 1980s and '90s. A discourse was taken to create its own object, and the media were deemed to be just one more industry, ruled by dominant economic forces, that diminished ideological or generic innovation and professionalism in favour of crass standardisation. Everything was distorted, and all knowledge was so contingent—even dubious—that it called faith in science into question the same way that rational people doubted religion.

Why was this so? Because although the signs we use to communicate had begun as reflections of reality, they'd been corporatized through the *bourgeois* media, such that two supposedly delineable things—truth versus lies—become indistinct. Reality was lost, signs were self-referential, and they lacked even a residual correspondence to the real. Signs had adopted the form of their own simulation—signifiers whose signified was themselves rather than a material referent (Baudrillard, 1988).

But this went further than the media. Supposedly compelling notions of truth, such as science, history, religion, and philosophy, had lost credibility due to their longstanding service in support of imperialism, slavery, colonialism, Maoism, misogyny, inequality, war, poverty, and fascism. In the process, established ideas of how to model truth, by creating something that corresponded through words or pictures to real life, or creating something that cohered as an argument, were fatally compromised (Malpas, 1992). Avowedly realistic war reportage, for example, borrowed from realism as a code of cinema and editorial norms of sensation and/or dependence on partisan sources as referents, as much as the events unfolding before it.

And today is the moment of 'fake news' and compromised binary notions of truth, as per US *bourgeois* journalism's characterization of all perspectives as divided between liberal and conservative *Weltanschauungen* (Ott, 2017). Environmental sociologists worry that the postmodern, which originated as a set of interesting intellectual, artistic, and architectural movements, has become accepted in a broader and deeper rejection of expertise (Lockie, 2017). That sense of despair reached its apogee in 2016, when:

> [a]fter much discussion, debate, and research, the Oxford Dictionaries Word of the Year 2016 is *post-truth*—an adjective defined as 'relating to or denoting circumstances in which objective facts are less influential in shaping public opinion than appeals to emotion and personal belief'.

But in the same year that the Oxford Dictionaries were forced to recognize this post-truth era, we entered what the world's scientific community announced as the Anthropocene—an epoch characterized by major geological and ecological changes wrought by human activity (Subcommission on Quaternary Stratigraphy, 2016).

Which Concept Will Win Out—Post-truth, or the Anthropocene?

Whilst there is necessarily some debate over climate change—its geography, history, speed, nature, and future—this is akin to debates over correlations between high consumption of alcohol and illness or leaping from tall buildings at a single bound and injury. In other words, the science is largely agreed, but there are always exceptions, complications, the need for additional research, and the emergence of new policy recommendations. That is how science works, through dual processes of accretion and attenuation.

Like scientists in general, climate scholars emphasize the need for patience in undertaking and understanding their work, which relies on the steady accumulation of sometimes paradoxical and even contradictory data. After all, climate is the average of weather. It is history (Chakrabarty, 2014).

But in addition to the complexity of ordinary research, climate scientists confront an additional obstacle: the systematic distortion of their work by the Anglo media. Wealthy polluters and their servants in public relations feast on occasional and entirely ordinary scholarly disagreements among climate experts, which are assiduously and duplicitously reported as proof of a putative underlying weakness in the evidence of climate change (Lewandowsky, Oreskes, Risbey, Newell, & Smithson, 2015, p. 2; Maxwell and Miller, 2016; Maxwell, Raundalen, & Vestberg, 2015). *Bourgeois* public discourse is dominated by ideologues and pundits for hire, and climate-change scientists are rarely invited to their party.

Clearly, the media have a role to play in such a contest of wills; this book testifies to the fact. But what role do the media themselves play—not in establishing, clarifying, or distorting the truth, but in creating climate change itself?

This chapter is concerned with how they help to constitute that reality—not discursively, but materially. For part of fake news and post-truth is the media's abject failure to understand, address, or discuss their own complicity in our current dilemma as environmental actors, alongside more obvious participants, such as the extractive and manufacturing industries. Debates about climate change may take place within newspaper pages or across communications screens and frequencies, but the notion that those very pages, screens, and frequencies are themselves contributors to the problem is largely beyond the ken of media narratives.

In what follows, we examine the centuries of pollution produced in the making of the media, from early print to late telephony. While allocating a number to their precise impact on climate change is implausible, the fact of their environmental despoliation is clear, and the data we have been able to marshal specific to coal-powered energy are particularly telling. The media produce climate change.

The History of Media Pollution

Since at least the 1300s, papermaking has required vast amounts of clean water, as both a power source and an ingredient. With each innovation in the chemical processing of fibre, effluents flowed into waterways, increasing their toxicity (Braudel, 1973, pp. 295–296; Innis, 2007, 150–151, Strasser, 1999). By the mid-1400s, Gutenberg's eponymous method of printing was using copper, lead, antimony, and tin. Lead is now recognized as lethal and the ink was composed of lampblack, turpentine, and boiled linseed oil—the first of which was harmful to the lungs and mucous membranes; the second to the nervous system, liver, and kidneys; and the third to the skin. New hazards in print shops laden with chemically-altered natural elements and process residues of fumes, dust, and heavy metals (Braudel, 1973, p. 296; Eisenstein, 1983, p. 13; Innis 2007, p. 165).

From the early 1800s, coal-burning, steam-powered presses multiplied the potential volume of printed pages by three to four times and added synthetic elements. Two interrelated environmental issues soon emerged: demand for wood pulp surpassed rags as the main source of fibre, and waste paper became a disposal problem. Things worsened with the advertising industry's need for 'substantial amounts of paper for labels and cardboard for packaging' (Burke, 1975, p. 175; Strasser, 1999, pp. 91–92). In the last quarter of the nineteenth century, new methods made it feasible to use wood fibre for industrial papermaking, initiating deforestation in the service of print technology and imposing grave risks on animal and plant diversity and habitats (Braudel, 1973; Innis, 2007).

Between 1899 and 1919, these and other innovations—such as illustrations and telegraphy—became standard. Tonnage consumption of processed wood pulp expanded by 1,175% in the US alone. Expenditure on print advertising grew 742% over the same period. Newspaper, book, and paperboard consumption escalated from 25 pounds per capita in 1909 to 59 pounds in 1930. In 1925, one New York newspaper accounted for 2,000 acres of forest (Burke, 1979, pp. 180–181, 188–190).

By 1930, papermaking had become 'one of the principal industries polluting water.' Throughout the twentieth century, print's rising energy demands increased the use of coal, and new chemical processes introduced large quantities of bleaching chemicals into the environment. This produced synthetic by-products such as dioxin. Despite new production techniques, environmental hazards for print workers remain serious. Today's pressrooms expose workers to airborne and liquid toxins, including solvents, developing solutions, and inks, at the same time as they emit lower atmosphere ozone (smog), heavy metals in ink, solvents, silver (film development), film and paper scraps, and wastewater into the environment. The chemicals released mainly come from solvents containing toluene, methyl ethyl ketone, xylene, and trichloroethane. Exposure to them can disrupt the normal functioning of internal organs and nervous systems in people and other animals (United States Department of Labor, n. d.).

Greenhouse gases emitted from paper and pulp makers include carbon monoxide, nitrogen dioxide, particulate matter, sulphur dioxide, and volatile organic compounds. In some cases, these substances are released at significantly higher levels than from electronics and computer manufacturing, and in certain categories, beyond even mining and petroleum (Independent Press Association, Conservatree, and Co-op America, 2001, p. 6; Organisation for Economic Co-Operation and Development, 2001, p. 18).

By 2001, the US produced nearly 100 million tons of paper annually. Approximately 90% was not recycled (Independent Press Association et al.,

2001, p. 6; Organisation for Economic Co-Operation and Development, 2001, p. 18). Today, just 20% of newspapers are recycled or end their time in landfills without even having been read, while each book produced in the US averages four kilograms of carbon dioxide emissions—eight times its average weight (Canonico, Royston, & Preist, 2009; Edwards, McKinnon, & Cullinane, 2009; Paper Task Force, 1995, p. 4; Sibley, 2009). By 2009, London's newspapers and books made up 13% of the city's consumer carbon footprint (BioRegional and London Sustainable Development Commission, 2009).

The chemical and paper industries continue to be the US's leading industrial users of water, with paper its fourth largest emitter of toxins into waterways and industrial consumer of energy. That helps make the industry the fourth biggest emitter of greenhouse gases. Add to this the toxic problems of dumping paper waste into landfills, carbon emissions from the transportation of printed material, and the industry's reliance on ecologically questionable monoculture plantation forestry for virgin fibre, and you get a stark picture of newsprint's negative environmental impact (Maxwell & Miller, 2012).

Publishers maintain that they're addressing the issue: paper producers increased recycling from 5% of all fibre in 2004 to 24% in 2010, reduced carbon emissions by 25% between 2006 and 2010, and save five million trees annually. They argue for their plantation program on the ground that youthful trees absorb carbon more readily than elderly ones, but just 10% of the paper used by publishers comes from recycled fibre (Book Industry Environmental Council and Green Press Initiative, 2013).

By contrast, electronic tablets provide some apparent benefits to consumers and the environment. The convenience of downloads saves transportation costs and reduces carbon emissions from newspaper distribution and consumer travel to and from newsstands. Research suggests that a hundred titles read on a tablet pays back the planet for the damage caused through the manufacture, use, and disposal of the device—mining, energy-intensive production, recharging, and end-of-life toxic waste—though industrial secrecy makes this more guesswork than certainty (Ritch, 2009).

That said, the environmental costs of producing a tablet (raw materials, transport, energy, and disposal) far outweigh those of one book printed on recycled paper: the e-reader uses 33 pounds of minerals, including tantalum, versus a paper book, which uses two-thirds of a pound; 79 gallons of water versus 2 gallons; and 100 kilowatt hours of fossil fuels versus two hours, with proportional emissions of carbon dioxide. In addition, the health effects of exposure to internal toxins is estimated to be seventy times greater for e-readers (Moberg, Johansson, Finnveden, & Jonsson, 2010).

These tablets are of course only the latest electronic media devices, whose story begins with practical applications of wireless communication that built upon experiments to describe, detect, and measure electromagnetic waves. The electromagnetic spectrum is comprised of ionizing radiation (ultraviolet rays, X-rays, gamma rays) and nonionizing radiation (extra-low and very low frequencies; electrical power lines, radio waves, microwaves). Radio and microwave frequencies are used in all the communications media that we are familiar with today—from garage-door openers and remote controls to TVs, cell phones, and satellite systems (Massey, 1979).

The proliferation of electronics in the post-World War II period prompted the US Department of Defense (DoD) to fund research into the dangers of radio-frequency radiation (RFR), especially radar and other microwave communications technologies. Because the DoD program focused only on the thermal effects of RFR, it created the perception among US radiation experts that non-thermal hazards from low-level RFR were non-existent. This perception continues to shape how regulators in the US respond to research that presents results showing stronger associations between harmful effects (cancer, behavioural and developmental problems, migraines …) and cell-phone use, wi-fi exposure, and other sources of non-thermal RFR (Massey, 1979).

Additional research has shown that the important parameters of exposure to RFR include the energy level of signal generation, proximity to the signal source, RFR resonance with affected bodies (absorption rates vary among species and body sizes), pulsed versus continuous wave forms, and duration of exposure. By 1980, there was consensus about a number of adverse effects associated with radiofrequency radiation: bio-thermal damage to organisms located in close proximity to transmission towers and signal generators, which included not only media workers continuously exposed to radio, TV, and telecommunication equipment, but also office workers on top floors of buildings within range of high-power transmission antennae (National Research Council, 2005, pp. 133–137).

In addition, with the advent of electronic television in the 1930s, the electron beams inside cathode-ray sets generated electromagnetic radiation in the X-ray portion of the spectrum, using high energy to excite cesium to create the electron cloud. The glass tubes containing the electron beams were designed to inhibit the escape of X-rays; the protective shields contained large quantities of lead in the glass (between four and eight pounds per monitor). Monochrome TVs were tested in the 1950s and 1960s for X-ray emission and found to be relatively safe. In contrast, the first generation of colour-television receivers leaked so much radiation that General Electric recalled 90,000 of them in 1967. The publicity about X-ray leakage from the sets helped generate regulatory protection (Cox, 2009; Electronics, 1967; Hayashi, Kitao, Hayashi, Oki, & Oshino, 1964; Massey, 1979).

When electronic-grade silicon crystal production was refined in 1962, the ecological context of media technologies changed once again. Toxic by-products began to flow into the environment from mining and chemical processing. High-grade silicon wafer production required 160 times more energy per kilogram than regular industrial-grade silicon. Mining raw silica, silicon, and ferrosilicon damages and pollutes the land, though it is believed to cause less harm to the environment than copper and gold mining because it doesn't employ chemical acids. Nevertheless, silicon dust can scar miners' lungs, inhibit breathing, and lead to incurable lung disease (silicosis) and susceptibility to tuberculosis and emphysema (Grossman, 2006, pp. 36–37).

Military specifications for microchips required not only that toxins be built into them, but that other known bio-toxins and pollutants be employed in manufacturing, emitting vast amounts of poisonous by-products. Only in the 1980s was the extent of pollution from US silicon manufacturers disclosed. By that time, hundreds of workers in Silicon Valley and people living close to semiconductor factories had developed cancer, reproductive disorders, and other illnesses (Grossman, 2006, pp. 37–41; Pellow & Park, 2002, p. 77).

Transmission and reception towers introduced environmental problems. They relate to the hazards encountered through RFR. In addition, the configuration of the towers, guy wires, and transmission cables brought about changes to land use, obstruction of the flight path of migratory birds, and aesthetic alteration of landscapes. Communications towers and wires may enable journalists to record and distribute their copy, but they also kill tens of millions of birds (over two hundred species) in North America every year, and an estimated 174 million annually across Europe and the US. The problem worsens with the proliferation of ever-taller towers (Kaufman, 2011; Rudolph & Swan, 1986).

Satellite communication, crucial to much international reportage from sports to wars, adds at least two toxic aspects via internal nuclear power sources and fuels, along with batteries, fuel cells, and solar panels; and basic rocketry, which emits messy discharges and scrap to the electronic waste stream. A vast array of 'space junk' has accumulated since the launch of the first satellite in 1957, most of it from the 1980s, when the satellite industry was deregulated, commercial satellite companies proliferated, and direct-satellite broadcasting began to win customers away from cable services. A 'debris belt,' predicted in the 1970s, now encircles the planet.

Meanwhile, the electronic media provide a complex collection of significant environmental problems—energy consumption and emissions, materials sourcing and toxicity, and electronic waste. There are between ten and fifteen billion high-tech devices needing electricity today, and 15% of global residential energy is spent powering domestic digital technology. If energy demand continues to grow at this rate, the residential electricity needed to power electronics will rise to 30% of global consumption by 2022, and 45% by 2030 (International Energy Agency, 2009). When residential use is added to the electricity it takes to make and distribute these goods, the total energy consumed translates into carbon emissions that are about the same as current levels from aviation—and this does not account for the energy to make chemicals and gases that go into the production of semiconductors or the energy used to dispose or recycle the devices.

As long as we don't see the smokestacks and pollution that accompany electricity needed for digitization, high-wattage operations of digital networks, office equipment, and video displays can be sold as clean, environmentally benign technologies ('Managing Energy', 2011). But these systems are plugged into the utility grid, which makes them part of climate change (International Energy Agency, 2009). The Carbon Trust has shown that people watching sports via mobile telephony multiply their footprint tenfold in comparison with television or wi-fi signals. The drain on power sources in other countries broadcasting the men's World Cup, for instance, is huge—and rarely acknowledged (National Grid, 2010, 2014).

And when cell phones are left in semi-repose, but still turned on, 400 terawatt hours per year are used annually, equivalent to 'the annual electricity generated by 133 mid-size coal-fired power plants (500 megawatts each), each requiring 1.4 million tonnes of coal per year' (International Energy Agency, 2014).

Then there is the electricity used in on-line searches and exchanges of information. Most searches are probably coal-powered, though their energy sources are as hard to pin down as the packet system itself. We know that Google's 2011 carbon footprint was almost equal to that of Laos or the United Nations Organization, largely due to running its search engines (Clark, 2011). Worldwide, there are now over fourteen billion networked media. They all need electricity to function, as does the system that enables enormous amounts of data to pass daily through massive networks and data centres—the 'cloud' (Carruth, 2014). Data centres' energy demands rise at a steady pace, with business practices that range from serious plans to reduce reliance on coal-fired energy to widespread examples of waste and thoughtless energy management. At current levels, cloud computing eats up energy at a rate somewhere between what Japan and India consume (Greenpeace International, 2012). The environmental impact of this networked culture depends on the type of energy production used to power the grids—coal-fired power being the biggest menace. The US National Mining Association and the American Coalition for Clean

Coal Electricity gleefully avow that the 'Cloud Begins with Coal' (Mills, 2013). The telecommunications industry provides network connectivity to mobile devices; it consumes up to 90% of the total energy needed to transfer digital content between journalists, editors, and readers (Center for Energy Efficient Telecommunications, 2013).

And when old and obsolete digital technologies are discarded, they become electronic or e-waste, which is the fastest-growing component of municipal clean-ups around the Global North. E-waste generates serious threats to workers' and residents' health and safety wherever plastics and wires are burnt, monitors smashed and dismantled, and circuit boards grilled or leached with acid, while the toxic chemicals, noxious gases, and heavy metals that flow from such practices have perilous implications for people, soil, and water both locally and downstream. Epidemiologists and other health scholars have shown that e-waste recyclers in the informal economy, ragpickers, suffer an historically unprecedented prevalence of low hemoglobin, high monocyte and eosinophil counts, gum disease, diarrhea, and dermatitis. Extraordinary levels of psychological distress are also reported (Devi, Swamy, & Krishna, 2014; Premalatha, Tabassum-Abbasi, Abbasi & Abbasi, 2014; Reis de Oliveira, Moura Bernardes, & Engel Gerbase, 2012).

What Is to Be Done?

As our brief historical account illustrates, both paper-based and electronic media have been major polluters for a very long time, through energy consumption and emissions, materials, toxicity, and waste. How might the media assess and mitigate their environmental impact, especially those charged with reporting on climate change to the citizenry?

The Society of Professional Journalists has made no discernible reference on line to the carbon footprint of its members since 2007. The National Union of Journalists adopted a policy calling for greener workplaces that same year, but shows no taste for problematizing its complicity in the problem. The International Federation of Journalists is silent on the topic. This *lacuna* is apparent even when organizations unfurl their credentials as committed to reporting climate-change science—the Project for Improved Environmental Coverage fails the test of reflexivity about its own carbon footprint. WWF India's otherwise excellent 'Recommendations on Environmental Journalism' (WWF, 2009) do not address the necessity of knowing and publishing the carbon footprint of reportage, from all moments in its life cycle—mineral extraction, manufacture, transport, and consumption. And the profession's codes of ethics need amending to address this failure.

The *Guardian* is one of the few newspapers that have both stimulated debate in this area (Dodd, 2007) and opened themselves up to public, scholarly scrutiny of their environmental impact (Wood, Shabajee, Schein, Hodgson, & Preist, 2014). In keeping with that example, journalists and news organizations must cultivate an eco-ethical mode of operation—a slow journalism (Le Masurier, 2016). That would help them come up with ways to account for the social liabilities of the media that we have outlined in this chapter—in particular, such negative environmental outcomes as harmful carbon emissions linked to the energy consumption and the waste associated with paper production, cloud computing, network operations, and electronics.

We therefore propose a form of very public green media accountancy (Greenham, 2010; Owen, Gray, & Bebbington, 1997) as per the UN's System of Environmental-Economic Accounting. This accountancy must include an element of public communication, such as guarantees at the end of motion pictures that no animals were harmed in the making of a film or audience information about violence on screen. But such initiatives need to transcend barely-legible medicinal warnings on packets and their cultural equivalents. We need the manufacturers of technologies and texts alike to emphasize the climatic impact of their work and its consumption (for instance, watching a football game on one's telephone) and post-consumption (what happens to that phone when it is declared obsolete and becomes e-waste).

We are standing over an abyss. Our climate is changing in ways that imperil us, our fellow animals, other forms of life, and the very Earth itself. Past and present industrial processes have exposed the planet to potentially irrevocable harm. Yet despite the evidence, the contours of this abyss are far from clear to many people.

The media contribute to this regrettable ignorance in a multitude of ways, as numerous studies, and other chapters in this volume, ably attest. One core component often forgotten or bypassed in this context is the polluting conduct of the media themselves, from journalists to readers, proprietors to regulators, editors to letter writers, and printers to cloud engineers. We hope this little chapter has shed specks of light on that impact.

References

Baudrillard, J. (1988). *Selected writings* (M. Poster, Ed.). Stanford: Stanford University Press.

BioRegional and London Sustainable Development Commission. (2009). *Capital consumption: The transition to sustainable consumption and production in London*. London: London Sustainable Development Commission.

Book Industry Environmental Council and Green Press Initiative. (2013). *Book industry environmental trends.* Retrieved from www.greenpressinitiative.org/documents/TrendsReport2013.pdf

Braudel, F. (1973). *Capitalism and material life 1400–1800* (M. Kochan, Trans.). New York, NY: Harper Calophon.

Burke, J. G. (1979). Wood pulp, water pollution, and advertising. *Technology and Culture, 20*(1), 175–195.

Canonico, S., Royston, S., & Preist, C. (2009). Reducing the greenhouse gas emissions of commercial print with digital technologies. *Proceedings of the 2009 IEEE International Symposium on Sustainable Systems and Technology.* Retrieved from http://www.hp.com/hpinfo/newsroom/press_kits/2009/ecosolutions/reduceimpact/ReducingGreenhouseGasEmissions.pdf

Carruth, A. (2014). The digital cloud and the micropolitics of energy. *Public Culture, 26*(2), 339–364.

Center for Energy Efficient Telecommunications. (2013). *The power of wireless cloud: An analysis of the energy consumption of wireless cloud.* Retrieved from http://www.ceet.unimelb.edu.au/pdfs/ceet_white_paper_wireless_cloud.pdf

Chakrabarty, D. (2014). Climate and capital: On conjoined histories. *Critical Inquiry, 41*(1), 1–23.

Clark, D. (2011, September 8). Google discloses carbon footprint for first time. *Guardian.* Retrieved from http://www.guardian.co.uk/environment/2011/sep/08/google-carbon-footprint

Cox, S. (2009). Cell phones generate particularly dangerous E-waste. In C. A. Bily (Ed.), *What is the Impact of E-Waste?* (pp. 18–26). Detroit: Greenhaven.

Devi, K. S., Swamy, A. V. V. S., & Krishna, R. H. (2014). Studies on the solid waste collection by rag pickers at Greater Hyderabad Municipal Corporation, India. *International Research Journal of Environment Sciences, 3*(1), 13–22.

Dodd, C. (2007, April 9). Carbon copy. *Guardian.* Retrieved from http://www.theguardian.com/environment/2007/apr/09/travelsenvironmentalimpact.mondaymediasection

Edwards, J. B., McKinnon, A. C. & Cullinane, S. L. (2009). *Comparing CO_2 emissions for conventional and online book retailing channels in the UK.* Edinburgh: Heriot-Watt University.

Eisenstein, E. L. (1983). *The printing revolution in early modern Europe.* Cambridge: Cambridge University Press.

Electronics: X rays in the living room. (1967, August 4). *Time.* Retrieved from http://www.time.com/time/magazine/article/0,9171,837185,00.html

Greenham, T. (2010). Green accounting: A conceptual framework. *International Journal of Green Economics, 4*(4), 333–345.

Greenpeace International. (2012). *How green is your cloud?* Retrieved from http://www.greenpeace.org/international/Global/international/publications/climate/2012/iCoal/HowCleanisYourCloud.pdf

Grossman, E. (2006). *High tech trash: Digital devices, hidden toxics, and human health.* Washington, DC: Island.

Hayashi, S., Kitao, T., Hayashi, M., Oki, I., & Oshino, A. (1964). X-rays from color television receivers. *Journal of Radiation Research, 5*(3), 147–158.

Independent Press Association, Conservatree, and Co-op America. (2001). *Turning the page: Environmental impacts of the magazine industry and recommendations for improvement.* Washington, DC: The Paper Project.

Innis, H. A. (2007). *Empire and communications.* Toronto: Dundurn.

International Energy Agency. (2009). *Gadgets and gigawatts: Policies for energy efficient electronics—Executive summary.* Paris: Organization for Economic Cooperation and Development.

International Energy Agency. (2014). *More data, less energy: Making network standby more efficient in billions of connected devices.* Retrieved from http://www.iea.org/publications/freepublications/publication/MoreData_LessEnergy.pdf

Kaufman, L. (2011, January 18). Conspiracies don't kill birds. People, however, do. *New York Times*, p. D4. Retrieved from https://www.nytimes.com/2011/01/18/science/18birds.html

Le Masurier, M. (2016). Slow journalism. *Journalism Practice, 10*(4), 439–447.

Lewandowsky, S., Oreskes, N., Risbey, J. S., Newell, B. R., & Smithson, M. (2015). Seepage: Climate change denial and its effect on the scientific community. *Global Environmental Change, 33*, 1–13.

Lockie, S. (2017). Post-truth politics and the social sciences. *Environmental Sociology, 3*(1). doi:org/10.1080/23251042.2016.1273444

Malpas, J. (1992). Retrieving truth: Modernism, post-modernism and the problem of truth. *Soundings, 75*(2–3), 288–306.

Managing energy costs in museums. (2011). *Energy Right.* Retrieved from http://www.energyright.com/business/pdf/Museums_ESCD.pdf

Massey, K. A. (1979). The challenge of nonionizing radiation: A proposal for legislation. *Duke Law Journal, 1*, 105–189.

Maxwell, R., & Miller, T. (2012). *Greening the media.* New York, NY: Oxford University Press.

Maxwell, R., & Miller, T. (2016). The propaganda machine behind the controversy over climate science: Can you spot the lie in this title? *American Behavioral Scientist, 60*(3), 288–304.

Maxwell, R., Raundalen, J., & Vestberg, N. L. (Eds.). (2015). *Media and the Ecological Crisis.* New York, NY: Routledge.

Mills, M. (2013). *The cloud begins with coal: Big data, big networks, big infrastructure, and big power: An overview of the electricity used by the global digital ecosystem.* National Mining Association/American Coalition for Clean Coal Electricity. Retrieved from http://www.americaspower.org/sites/default/files/Cloud_Begins_With_Coal_Exec_Sum.pdf

Moberg, Å., Johansson, M., Finnveden, G., & Jonsson, A. (2010). Printed and E-tablet newspaper from an environmental perspective—A screening life cycle assessment. *Environmental Impact Assessment Review, 30*(3), 177–191.

National Grid. (2010). *National grid powers up for World Cup 2010.* Retrieved from http://www.nationalgrid.com/uk/Media+Centre/WorldCup2010

National Grid. (2014, May 30). *National grid up for the World Cup.* Retrieved from http://www2.nationalgrid.com/mediacentral/uk-press-releases/2014/national-grid-up-for-the-world-cup/

National Research Council. (2005). *An assessment of potential health effects from exposure to PAVE PAWS low-level phased array radiofrequency energy.* Washington, DC: National Academies.

Organisation for Economic Co-Operation and Development. (2001). *Environmental outlook.* Paris: Organisation for Economic Co-Operation and Development.

Ott, B. (2017). The age of twitter: Donald J. Trump and the politics of debasement. *Critical Studies in Mass Communication, 34*(1), 59–68.

Owen, D., Gray, R., & Bebbington, J. (1997). Green accounting: Cosmetic irrelevance or radical agenda for change? *Asia-Pacific Journal of Accounting, 4*(2), 175–198.

Paper Task Force. (1995). *Paper task force recommendations for purchasing and using environmentally preferable paper: Project synopsis.* New York, NY: Environmental Defense Fund.

Pellow, D. P., & Park, L. S. (2002). *The silicon valley of dreams: Environmental justice, immigrant workers, and the high-tech global economy.* New York, NY: New York University Press.

Premalatha, M., Tabassum-Abbasi, Abbasi, T., & Abbasi, S. A. (2014). The generation, impact, and management of E-waste: State of the art. *Critical Reviews in Environmental Science and Technology, 44*(14), 1577–1678.

Reis de Oliveira, C., Moura Bernardes, A., & Engel Gerbase, A. (2012). Collection and recycling of electronic scrap: A worldwide overview and comparison with the Brazilian situation. *Waste Management, 32*(8), 1592–1610.

Ritch, E. (2009). *The environmental impact of Amazon's Kindle.* San Francisco, CA: CleanTech Group.

Rudolph, L., & Swan, S. H. (1986). Reproductive hazards in the microelectronics industry. *Occupational Medicine, 1*(1), 135–143.

Sibley, L. (2009, August 19). Cleantech group report: E-readers a win for carbon emissions. *Cleantech.* Retrieved from http://www.cleantech.com/

Strasser, S. (1999). *Waste and want: A social history of trash.* New York, NY: Metropolitan.

Subcommission on Quaternary Stratigraphy, Working Group on the 'Anthropocene.' (2016). *What is the 'Anthropocene'?: Current definition and status.* Retrieved from http://quaternary.stratigraphy.org/workinggroups/anthropocene/.

United States Department of Labor. (n. d.). US Occupational Safety and Health Administration. *Printing Industry.* Retrieved from https://www.osha.gov/SLTC/printing_industry/index.html.

Wood, S., Shabajee, P., Schein, D., Hodgson, C., & Preist, C. (2014). Energy use and greenhouse gas emissions in digital news media: Ethical implications for journalists and media organisations. *Digital Journalism, 2*(3), 284–295.

WWF. (2009, March 11). Recommendations on environmental journalism. Retrieved from http://wwf.panda.org/?158643/Recommendations-on-environmental-journalism

11. *The Politics of Place: Networking Resistance to Coal Seam Gas Mining*

ALANA MANN

Introduction

The hydraulic fracturing or fracking of coal seam gas (CSG) has increased with the decline of natural gas reserves from conventional sources, increased export demand, and the search for clean energy to replace fossil fuels. CSG is promoted by the Australian Government as one of the transition fuels that will lead the country toward a 'clean energy future' while providing 'green jobs', enhancing national security, and stabilising energy prices (Australian Government, 2012). However the siting of CSG activities on productive agricultural land has led to the emergence of political communities concerned with the impact of fracking on the environment, local livelihoods, and national food security.

This chapter interrogates how land use changes trigger people to express their attachment to place and motivate them to become politically active. It does so through analysis of competing problem definitions or frames presented by politicians, activists, and mining companies to better understand the political dynamics and conditions under which oppositional publics emerge and form discourse coalitions (Hajer, 1993) using traditional and digital methods of organising. The focus of the study is a grassroots social movement of over 250 groups in Australia and New Zealand, the Lock the Gate Alliance (LTGA). This movement is credited with uniting "cockies, blockies, croppers and greenies" (Hutton, 2012) in non-cooperation campaigns against mining to protect water systems, agricultural land, bushlands, wetlands and wildlife, health, and Aboriginal and cultural heritage. Effective information-sharing, mobilisation of members and a high degree of interconnectedness between advocacy groups has contributed to the impact of direct actions against Australian projects such as the proposed Shenhua Watermark coal mine on the Liverpool Plains, CSG extraction in the Northern Rivers region, and the fracking of tight gas in the Kimberley region. Characterised by a network-centric approach that includes tactics such as grassroots lobbying, physical protests and digital activism, the LTGA has been influential in winning support for its campaign to drive extractive industries off productive agricultural land.

'Clean' Energy and the Emergence of an Oppositional Discourse

Uncertainty and dysfunction characterise energy policy in Australia. The Liberal Government has limited any action on climate change to unambitious election commitments, with Prime Minister Malcolm Turnbull warning against a 'mindless rush into renewables' (Talberg & Workman, 2016). In his January 2017 Press Club speech Turnbull said that priority would be given to "clean coal and carbon capture and storage (CCS) and onshore gas (CSG)" and went on to declare that Australian energy policy should be "technology agnostic—it's security and cost that matter, not how you deliver it. Policy should be 'all of the above technologies' working together to meet the trifecta of secure and affordable power while meeting our (substantial) emission reduction commitments". (Turnbull, 2017)

The Prime Minister's claim that 'increasing gas supply in Australia is vital for our energy future and vital for industries and jobs' extends to farmers who require reliable, affordable and secure energy to grow food. These needs are likely to increase as our food systems reveal their extreme vulnerability to climate change. Rising temperatures and extreme weather events are likely to irreversibly damage the natural resource base on which agriculture and the livelihoods of food producers depend world-wide (FAO, 2017). In all countries the price of food is expected to rise dramatically while the quality and seasonal availability of produce will be reduced. Australia's *National Climate Resilience and Adaptation Strategy* (Australian Government, 2015) sets as Priority 1 the need to 'understand and communicate' the risks of a changing climate, which include increased frequency of drought conditions in southern Australia, extreme heat events, rising water temperatures, changes in growing seasons, and biosecurity threats including pests, diseases, and weeds. Climate change in other parts of the world could influence Australian farmers via changes in market prices (FAO, 2007).

These concerns are neglected in government discourse which is market-driven and dominated by contestation over appropriate regulation rather that the threat of mining to food and water supplies. The National Partnership Agreement for the Regulation of Coal Seam Gas was developed in 2012 to inform regulation with 'strong science' (Bubna-Litic, 2015, p. 439) in clear alignment with the Australian Petroleum Production and Exploration Association (APPEA). Both parties emphasise the contribution of CSG to regional development through foreign investment, economic growth, and infrastructure (Mercer, de Rijke, & Dressler, 2014). In its role as the 'voice of oil and gas in Australia', APPEA reports:

> CSG can play an important role in regional development. The industry is bringing infrastructure and investment to several rural and regional districts, providing new jobs and strengthening and diversifying regional economies. (APPEA, 2013)

The Association promotes CSG as the 'pathway to net-zero emissions' claiming that 'more gas, not less, is the way to reduce emissions' (APPEA, 2017). Meanwhile, the contribution of fracking to emissions remains contested. In the absence of independent studies in Australia, US studies of so-called 'unconventional' gas projects have been drawn on to estimate fugitive emissions. Fugitive emissions, a byproduct of the coal seam gas mining process, are comprised of methane, a greenhouse gas 'more than 100 times more powerful than carbon dioxide' but with a much shorter life in the atmosphere (Howarth, 2015, p. 46). While companies talk down emissions from CSG, LTGA points to evidence that these emissions make CGS 'potentially more polluting than coal when burnt for electricity' (Lock the Gate, 2017a). It recommends, in line with the Melbourne Energy Institute's (2016) report, that an independent regulatory body, funded by a levy on the industry, be created and that emissions monitoring be conducted and results made public. In response APPEA claims that LTGA seeks 'to mislead regional communities and the media by misrepresenting important research' (APPEA, 2016).

LTGA challenges the legitimacy of Government and industry claims based on insufficient data, stating

> We believe that neither we, nor our governments (at all levels), are sufficiently well informed about these industries, about their true role in our economy and their impact on our health and welfare as a nation and within the diverse communities that constitute our nation. (LTGA Aotearoa, 2012)

Focusing on community values, ecological and personal health, and the irreplaceable nature of rural lifestyles, the LTGA is largely silent on local economies and jobs. This 'politics of silence' (Mercer et al., 2014, p. 294) is a symptom of the challenge members of the LTGA face in presenting alternatives to regional development which, in Australia, is traditionally based on resource extraction i.e. Big Coal (Pearse et al., 2013).

Movement Building Through Communication

Resistance to novel forms of energy extraction is shared by members of what was become known as the transnational Blockadia movement. This 'grassroots and broad-based network the likes of which the environmental movement has rarely seen' (Klein, 2014, p. 295) opposes mining of oil sands (and the TransCanada's Keystone XL pipeline) and coal seam gas. Aided by the emergence of a growing sphere of social media activism, local campaigners are creating new opportunities for the scaling-up of concerns and dissent through transnational advocacy networks (Keck & Sikkink, 1998) while participating in local, place-based mobilisations (Cottle & Lester, 2011; Mann, 2014). These global discourse coalitions and networks are driven by a common sense of disillusionment with the ability of governments, Big Green and UN Summits to formally address wicked problems such as climate change. In this context defending local landscapes is a way of representing global stakes and demonstrating a desire for deeper democracy.

> Having the ability to defend one's community's water source from danger seems to a great many people like the very essence of self-determination. What is democracy if it doesn't encompass the capacity to decide, collectively, to protect something that no one can live without? (Klein, 2014, p. 361)

As a broad-based alliance of people from all backgrounds and ideological leanings, LTGA demonstrates features of this new way of doing politics. The 'locked gate' itself telegraphs powerful values and beliefs that include defense of country and rural livelihoods.

> A locked gate in the bush is a powerful symbol. It means you can't come in and it expresses the determination of landowners to refuse entry to resource companies even though the law might say they have no choice. (Hutton, 2012, p. 17)

LTGA's growth has been attributed to face-to-face community building to establish ownership of campaigns; training to participants in strategic nonviolent direct action such as blockades and networked framing, a 'process through which particular problem definitions, causal interpretations, moral evaluations, and/or treatment recommendations attain prominence through crowdsourcing practices' (Papacharissi, 2015, p. 75). Protests at Tara, Glenugie, Doubtful Creek and Fullerton Cove in Australia, and Taranaki and Tararua in New Zealand's East Coast Basin have created a groundswell of support and raised awareness through traditional and digital media channels. On Facebook opposition to CSG is voiced by more than 20 eastern Australian anti-CSG groups including CSG-Rescue NSW, CSG Free Gippsland, Youth Against

CSG, and Stop CSG Sydney Water Catchment. Slogans such as 'Australians uniting to protect our land and water', 'Protect our national foodbowl' and 'Water not Coal' capture common concerns. Photos and information about events including towns declaring themselves coal and gas-free, 'human' signs spelling out 'Stop CSG!' and abseilers dropping banners for cliffs are carefully selected and uploaded to social media platforms in an active process of audiencing material from protests that is then amplified by mainstream journalists and sympathetic politicians to the wider public. Hashtags such as #noCSG, #Water4Life, #KeepthePressureUp, #FrackFreeWA, and #csgcon operate beyond their role as 'lightweight, semantic annotations that publics assign to tweets in their efforts to self-tag generated content' to serve as framing devices (Meraz & Papacharissi, 2016, p. 104; Mann, 2018). Tweet frequency, based on hashtags including #fracking #CSG, #lockthegate, and #gasland, has been identified as a proxy for more traditional offline protest events at physical project sites and centers of power such as Parliament House (Kuch & Titus, 2014). Rich, shareable video content including geo-mapping of the Narrabri Santos mine site; live footage of the Condamine River, lit by surface gas; and cheeky memes featuring Prime Minister Malcolm Turnbull juxtaposed with the statement 'Don't Frack Up Australia!' frame calls to action (LTGA, 2017b). The personal stories of affected land-holders, sit-ins and testimonies from grassroots groups ranging from the County Womens' Association (CWA) to the Knitting Nannas present the lived experience of the conflict between energy and food production in rural Australia. Symbiotic, conversational interactions on topics of mutual interest between these groups on social media platforms represent new processes of frame negotiation that permit individuals to 'organise social and political meaning around their own lifestyle values' (Bennett, 2004) in a form of personally defined and expressed citizenship while still engaging in collective action driven by the logic of 'connective action' (Bennett & Segerberg, 2012). Connective action embodies self-motivated participation through the sharing of personal content that may be co-produced and co-distributed among peers with congruent ideas, plans, resources, and networks (Benkler, 2006). This pragmatic approaches to coalition building helps to unite diverse individuals in 'asserting their rights to contribute to their communities and participate in decisions that impact them in a meaningful way' (Paris, Ramos, & Pritchard, n.d., p. 3).

By framing extractive industries as a threat to national food security and rural life the movement has emerged as a real challenger to mining companies, in many cases thwarting their efforts to acquire social license to operate (Lacey & Lamont, 2014). A novel and increasingly successful strategy to achieve this is being conducted by the internationally networked fossil fuel divestment movement. Calling for public and private institutions to abandon their financial holdings in traditional energy companies, the divestment campaign has exposed the cracks in the business model of carbon extraction. Risk articulation and ethical investment campaigns are being recognised by corporations as causing not only disruption on the ground but ongoing reputational damage. In Australia, the LTGA is among those groups collaborating with Greenpeace, 350.org and Carbon Tracker to 'join the dots' for investors—including employee shareholders—through clearly articulating demonstrable risk scenarios (Ricketts, 2013). In aligning with the networked framing of the global divestment movement, members of LTGA join international activists in demanding 'forms of respect, forms of democracy and changes in the ways in which market processes are embedded in social life' (Bebbington, 2014, p. 34).

Framing Place Identity

The top-down 'conceive-decide-implement' model employed by the Australian Government in rolling out CSG ignores the importance of involving residents and other stakeholders in planning processes. Lack of consultation seriously erodes the opportunities for success of any place-based development project reliant on 'the creation of relationships with host communities and countries that are characterised by authenticity, respect, integrity, inclusiveness and transparency' (Hodge, 2014, p. 27). Successful community led technology initiatives in countries such as Germany demonstrate how participative and bottom-up planning processes are essential to achieve acceptance and cooperation on the basis of shared values and common interests (Li, Birmele, Schaich, & Konold, 2013). In Friemt, in the Black Forest, farmers were the original drivers of renewable energy development. The Federal Republic of Germany's Renewable Energy Act supports small-scale development of renewable energy, and counts residents and local government as key stakeholders. In contrast, the CSG debate highlights the power differentials between scientific and governing elites and local communities that constrain equitable decision-making (Armitage et al., 2008). Local residents apply cultural rationality that gives equal or greater weight to personal, lived experience than 'depersonalised technical calculations' (Fisher, 2000), yet the knowledge cultures of local communities campaigning against CSG are commonly framed by bodies such as APPEA as lacking credibility in comparison to elite discourses (Brown, 2008).

Despite the existence of guidelines for community consultation on CSG mining in the state of New South Wales (2012) policy does not appear to reliably represent the feelings of various constituencies, given their contesting framings of the issue. Framing is a critical variable in energy policy emergence and evolution where framings constitute 'storylines or narratives about the problem of energy system change; how it has arisen, why it matters, and what to do about it' (Butler, 2015, p. 667). Frame analysis explores how patterns of discourse are established and how they mobilise publics. The verification of frames is made through the close study of campaign discourse, the 'stuff of framing' (Steinberg, 1998, p. 862). This 'stuff' includes layers of discourse: the storylines, myths, and metaphors such as the 'locked gate', policy vocabularies supported by expert discourses and epistemic figures such as 'transition fuels' and 'clean energy'. Erving Goffman (1974) first applied the concept of framing in studies of communicative acts, claiming that 'to frame is to select some aspects of a perceived reality and make them more salient in a communicating text, in such a way as to promote a particular problem, definition, causal interpretation, moral evaluation, and/or treatment recommendation' (Entman, 1993, p. 52). 'Issue frames' are deployed to shape the public's understanding of how a problem came to be while introducing important criteria by which solutions should be evaluated. This alignment of the individual frames, or 'schemata of interpretation' (Goffman, 1974), becomes the essence or *raison d'etre* of a social movement such as the LTGA. Prognosis requires agreement on what can be done to solve the problem that has been diagnosed, building an interpretive foundation that forms the basis of recruitment, participation, and action. Finally, frame resonance is created through the strategic communication of the movement ideology. It may require compromise to attract large numbers of supporters or influential actors to the movement. Congruence with the experience, attitudes, culture, and beliefs of the target public is essential if the frame is to mobilise action and therefore become motivational (Benford, 1993, p. 677; see also Entman, 1993).

Those who live in communities targeted for CSG extraction interpret exclusive reliance on 'expert' industry and government knowledges as denial of what is meaningful in the lived environment, and as indicative of a lack of appreciation for labour of previous generations and a way of life. For community members the landscape 'is not a mere 'surface' waiting for a new plan; the landscape was loaded with meanings and signifiers, stories and achievements' (Hajer & Wagenaar, 2003, p. 93). An anthropological study of the Darling Downs region of Queensland, site of the first LTGA office, identified recurring themes that concern food producers in the region including 'rural social relationships, emplaced identity and the impersonal imposition of industrial resource exploitation, clashing forms of environmental knowledge, the distribution and exercise of rights, as well as anxiety inducing transformations of material surroundings' (de Rijke, 2013, p. 46). As the study's author notes, capital, through technology, has already transformed food production in the Darling Downs. Today, Australia's 'agri-gas fields' are the sites of more radical transformations, leading to intensified conflicts over land, water, and food. Personal subjectivities are largely elided in these transformations, made in the name of capital growth.

How industry and policy-makers deal with the attachment of people to place is vital to the success of energy projects. In his case study of offshore wind farms in North Wales, Patrick Devine-Wright (2011) interviewed members of the action group Save our Scenery. While members were concerned with the threat the wind farm posed to the character of the area, developers and government framed the project in terms of tackling global warming and meeting national energy targets. The elites' collective failure to construct a narrative about the 'fit' between place and project is highlighted by Devine-Wright as 'lost opportunities to embed the project in the symbolic meanings of the place' (2011, p. 66). He notes that dominant framings concerning the 'where' of energy projects typically simplify local resistance as selfish NIMBYism ('not in my backyard') or as 'siting'. Siting draws attention to objective features, playing down the subjective, the 'memories, experiences, stories and myths that are as much a feature of any locality as the soil type, height above sea level or average wind speed' (Devine-Wright, 2011, p. 59). Devine-Wright suggests concepts of place attachment (Manzo, 2005) and place identity (Carrus, Bonaiuto, & Bonnes, 2005; Cresswell, 2013) that recognise and respect the emotional bonds and connections between people and their environments serve as more persuasive frames that arguments about climate change, and even affordable energy, in debates over land use.

Conclusion

Critics of Australia's 'clean coal' agenda define it as a war on renewables and accuse the government of attempting to mislead the public regarding the contribution of coal seam gas mining to harmful emissions *(Climate Council, 2017; Dunlop, 2017;* Lavoipierre, 2017*)*. Comparative analysis of the discursive frames produced by actors on each side of the CGS debate identifies linguistic elements that reduce complexity and marginalise alternatives, along with competing frames that open up possibilities.

In analysing how LTGA challenges state-sanctioned framings of productive resources such as land in terms of economic worth alone. This study contributes to the growing body of research on community social movement activism against CSG in Australia (de Rijke, 2013; Greer, Tabert, & Lockie,

2011; Kuch & Titus, 2014; Lloyd, Luke, & Boyd, 2013; Mercer et al., 2014; Ricketts, 2013). However it also offers a new perspective on how the digital media practices of activists unite them in 'imagined collective[s] that emerge as a result of the intersection of people, technology and practice' (Boyd, 2010, p. 39) through processes of networked framing. Using digital media to mobilise protestors and contest the role of CSG in Australia's energy policy the movement links the online and offline worlds of protest on multiple scales. These networked publics seek to determine *for themselves* what is a legitimate use of rural space, and particularly valuable food-producing land. Their resistance to the transformations imposed on them reflects a widespread desire for the reinvention of the politics of place, an aim challenged in a policy environment rife with the competing tensions of domesticating global emissions targets, providing access to affordable energy and respecting rural landscapes and lifestyles.

References

APPEA. (2013). Regional development. Retrieved January 10, 2016, from http://www.appea.com.au/oil-gas-explained/benefits/regional-development/

APPEA. (2016). LTGA misrepresents important CSG research. Retrieved January 10, 2016, from http://www.appea.com.au/media_release/lock-the-gate-misrepresents-important-csg-research/

APPEA. (2017). More gas, not less, is the way to reduce emissions. Retrieved April 20, 2017, from https://www.appea.com.au/media_release/use-more-gas-to-reduce-emissions/

Armitage, D., M. Marschke, and R. Plummer. 2008. Adaptive co-management and the paradox of learning. *Global Environmental Change*. 18:86–98.

Australian Government. (2012). *Australia's clean energy future plan: overview and opportunities.* Retrieved on February 28, 2017, from file:///C:/Users/Dave/Downloads/Clean%20Energy%20Future%20Plan%20DRET.pdf

Australian Government. (2015). *National climate resilience and adaptation strategy.* Retrieved on March 15, 2017, from http://www.environment.gov.au/climate-change/adaptation/strategy

Bebbington, A. (2014). Socio-environmental conflict: An opportunity for mining companies. *Journal of Cleaner Energy Production, 84*, 34.

Benford, R. D. (1993). Frame disputes within the nuclear disarmament movement. *Social Forces, 71*(3), 677–701.

Benkler, Y. (2006). *The wealth of nations: How social production transforms markets and freedom.* New Haven, CT: Yale University Press.

Bennett, W. L. (2004). Branded political communication: Lifestyle politics, logo campaigns, and the rise of global citizenship. In M. Micheletti, A. Follesdal, & D. Stolle

(Eds.), *Politics, products and markets: Exploring political consumerism, past and present* (pp. 101–125). New Brunswick, NJ: Transaction.

Bennett, W. L., & Segerberg, A. (2012). The logic of connective action. *Information, Communication & Society, 1*(5), 739–768.

Boyd, D. (2010). Social network sites as networked publics: Affordances, dynamics and implications. In Z. Papacharissi (Ed.), *Networked self: Identity, community, and culture on social network sites* (pp. 39–58). New York, NY: Routledge.

Brown, V. A. (2008). *Leonardo's vision: A guide to collective thinking and action.* The Netherlands: Sense Publishers.

Bubna-Litic, K. (2015). Fracking in Australia: The future in South Australia? *Environmental and Planning Law Journal,* 437–54.

Butler, J. R. A., Wise, R. M., Skewes, T. D., Bohensky, E. L., Peterson, N. Suadnya, W., Yanuartati, Y. Handayani, T., Habibi, P., Puspadi, K., Bou, N., Vaghelo D. & Rochester, W. (2015) Integrating Top-Down and Bottom-Up Adaptation Planning to Build Adaptive Capacity: A Structured Learning Approach. *Coastal Management.* 43:4, 346–364.

Carrus, G., Bonaiuto, M., & Bonnes, M. (2005). Environmental concern, regional identity and support for protected areas in Italy. *Environment and Behaviour, 37*(2), pp. 237–257.

Climate Council. (2017). Turbull's speech on energy is misleading. Retrieved March 7, 2017, from https://www.climatecouncil.org.au/turnbulls-speech-on-energy-is-misleading (accessed).

Cottle, S., & Lester, L. (2011). *Transnational protest and the media.* New York, NY: Peter Lang.

Cresswell, T. (2013). *Place: A short introduction.* Oxford: Blackwell.

de Rijke, K. (2013). The Agri-gas fields of Australia: Black soil, food and unconventional gas. *Culture, Agriculture, Food and Environment, 35*(I), 41–53.

Devine-Wright, P. (Ed.). (2011). *Renewable energy and the public: From NIMBY to participation.* London: Earthscan.

Dunlop, I. (2017). 'Clean coal', CCS and CSG will not save fossil fuels—Their game is up. *Guardian.* Retrieved July 7, 2017, from https://www.theguardian.com/sustainable-business/2017/mar/03/clean-coal-ccs-and-csg-will-not-save-fossil-fuels-their-game-is-up

Entman, R. M. (1993). Framing: Towards clarification of a fractured paradigm. *Journal of Communication, 43*(4), 51–58.

FAO. (2017). *The future of food and agriculture: Trends and challenges.* Retrieved April 13, 2017, from http://www.fao.org/3/a-i6583e.pdf

Goffman, E. (1974). *Frame analysis: An essay on the organisation of experience.* Cambridge, MA: Harvard University Press.

Fischer, F. (2000). *Citizens, experts, and the environment.* Durham: Duke University Press.

Greer, L., Tabert, S., & Lockie, S. (2011). *Food, coal or gas? Community action, land-use conflict and procedural fairness in the Surat Basin, Queensland.* Paper presented to the

Agri-Food XVIII: Sustainabilities, justice and agriculture in the Asia-Pacific Region Conference, Australian National University, Canberra, 5–8 December.

Hajer, M. (1993). Discourse coalitions and the institutionalisation of practice: The case of acid rain in Great Britain. In F. Fischer & J. Forester (Eds.), *The argumentative turn in policy analysis and planning* (pp. 43–67). Durham, NC: Duke University Press.

Hajer, M., & Wagenaar, H. (2003). *Deliberative policy analysis: Understanding governance in the network society.* Cambridge: Cambridge University Press.

Hodge, R. A. (2014). Mining company performance and community conflict: Moving beyond a seeming paradox. *Journal of Cleaner Production, 84*, 27–33.

Howarth, R. W. (2015). Methane emissions and climatic warming risk from hydraulic fracturing and shale gas development: Implications for policy. *Energy and Emission Control Technologies, 3*, 45–54.

Hutton, D. (2012). LTGA unites cockies, blockies, croppers and greenies. *Chain Reaction, 15*(August), 16–17.

Keck, M. E. & Sikkink, K. (1998). *Activists beyond borders: Advocacy networks in international politics.* Ithaca: Cornell University Press.

Klein, N. (2014). *This changes everything: Capitalism versus the climate.* New York, NY: Simon & Schuster.

Kuch, D., & Titus, A. (2014). Emerging dimensions of networked energy citizenship: The case of coal seam gas mobilisation in Australia. *Communication, Politics & Culture, 47*(2), 35–59.

Lacey, J., & Lamont, J. (2014). Using social contract to inform social licence to operate: an application in the Australian coal seam gas industry. *Journal of Cleaner Production, 84*, 831–839.

Lavoipierre, A. (2017, July 7). Federal government releases secret emissions data following reports it kept it hidden. *ABC News.* Retrieved from July 7, 2017, http://www.abc.net.au/news/2017-07-07/government-releases-secret-emissions-data-following-reports/8688584

Li, W. L., Birmele, J., Schaich, H., & Konold, W. (2013). Transitioning to community-owned renewable energy: Lessons from Germany. *Procedia Environmental Sciences, 17*, 719–728.

Lloyd, D. J., Luke, H., & Boyd, W. E. (2013). Community perspectives of natural resources extraction: Coal-seam gas mining and social identity in Eastern Australia. *Coolabah, 10*, 144–164.

LTGA. (2017a). Unconventional gas facts: Methane leakage and venting. Retrieved September 15, 2017, from http://www.lockthegate.org.au/methane_leakage_and_venting

LTGA. (2017b). *The Lock the Gate Alliance* [facebook page]. Retrieved April 20, 2017, from https://www.facebook.com/Lock.The.Gate.Alliance/

LTGA Aotearoa. (2012). Retrieved November 15, 2015, from http://www.lockthegate.org.nz/lock-the-gate-alliance.html

Mann, A. (2014). *Global activism in food politics: Power shift.* Basingstoke, Hampshire: Palgrave Macmillan.

Mann, A. (2018). Hashtag activism and the right to food in Australia. In T. Schneider, K. Eli, C. Dolan, & S. Ulijaszek (Eds.), *Digital food activism*. London: Routledge, pp. 168–84.

Manzo, L. (2005). For better or for worse: Exploring multiple dimensions of place meaning, *Journal of Environmental Psychology, 25*, 67–86.

Melbourne Energy Institute. (2016). A review of current and future methane emissions from Australian unconventional oil and gas production. Retrieved September 30, 2017, from http://energy.unimelb.edu.au/__data/assets/pdf_file/0019/2136223/MEI-Review-of-Methane-Emissions-26-October-2016.pdf

Meraz, S., & Papacharissi, Z. (2016). Networked framing and gatekeeping. In T. Witschge, C. W. Anderson, D. Domingo, & A. Herminda (Eds.), *The SAGE handbook of digital journalism* (pp. 95–112). SAGE: London.

Mercer, A., de Rijke, K., & Dressler, W. (2014). Silences in the book: Coal seam gas, neoliberalising discourse, and the future of regional Australia. *Journal of Political Ecology, 21*, 222–348.

Papacharissi, Z. (2015). *Affective publics: Sentiment, technology, and politics*. New York, NY: Oxford University Press.

Paris, N., Ramos, J., & Pritchard, M. (n.d). *Creating change in Australia*. Retrieved January 8, 2016, from http://counteract.org.au/wp-content/uploads/2013/11/creating-change-case-study-csg.pdf

Pearse, G., McKnight, D., & Burton, B. (2013). *Big Coal: Australia's Dirtiest Habit*. Sydney: New South,

Ricketts, A. (2013). Investment risk: An amplification tool for social movement campaigns globally and locally. *Journal of Economic and Social Policy, 15*(3), Article 4.

Steinberg, M. (1998). Tilting the frame: Considerations on collective action framing from a discursive turn. *Theory and Society, 27*(6), 845–872.

Talberg, A., & Workman, A. (2016). *Timeline: Australian climate and clean air policy interventions 2013–16: Developments under the Abbot-Turnbull Government*. Australian-German Climate & Energy College, University of Melbourne.

Turnbull, M. (2017, February 1). Address at the National press club. Retrieved from https://www.malcolmturnbull.com.au/media/address-at-the-national-press-club-and-qa-canberra

Contributors

Benedetta Brevini is Senior Lecturer in Communication and Media at the University of Sydney and Visiting Fellow of the Centre for Law, Justice and Journalism at City University, London. Dr Brevini is also an experienced journalist who has worked in Milan, New York, and London for CNBC and RAI. She writes on *The Guardian*'s Comment Is Free and contributes to a number of print and web publications including the Index of Censorship and OpenDemocracy and the Conversation. She is the author of *Public Service Broadcasting online* (2013) and editor of the acclaimed volume *Beyond Wikileaks* (2013). Her latest book is *Carbon Capitalism and Communication: Confronting Climate Crisis* (Palgrave Macmillan, 2017).

Stuart Capstick is an environmental social scientist interested in the ways in which people understand and respond to climate change. His research has considered how best to communicate environmental topics, how public attitudes towards climate change develop over time, people's interpretation of extreme weather events, cross-cultural dimensions of public perceptions and environmentally-significant behaviour, and approaches to promoting radical emissions reduction through lifestyle change. Stuart is based in the School of Psychology at Cardiff University and is affiliated with the Tyndall Centre for Climate Change Research.

Anabela Carvalho is Associate Professor at the Department of Communication Sciences of the University of Minho. Her research focuses on various forms of environment, science, and political communication with a particular emphasis on mediations of climate change, as well as connections to public engagement. Anabela Carvalho has been on the Board of Directors of the International Environmental Communication Association (IECA) (2011–2015), was co-founder, former Chair, and Vice-Chair of the Science and Environment Communication Section of the European Communication and Education Association (ECREA) (2007–2014), Associate Editor of the journal *Environmental Communication* (2010–2012), and is currently Associate Editor of *Frontiers in Science and Environment Communication*.

Sibo Chen is SSHRC Vanier Scholar in the School of Communication at Simon Fraser University. His research interests embrace environmental communication, critical discourse analysis, political economy, and energy-society relationship.

Victoria Dando is Deputy Head of Communication at Cardiff University. Her PhD examined the discursive and ideological conditions involved in the construction of climate change and terrorism in five British newspapers. Analysing 1,013 newspaper articles between 1999 and 2012, her research provided the first known comparison of the reporting of climate change and terrorism in the UK press.

Neil T. Gavin is a Senior Lecturer in the Department of Politics at the University of Liverpool. He has written extensively on the mediation of climate change for journals such as *Environmental Politics*, *Environmental Communication*, *British Journal of Politics and International Relations*, *Public Understanding of Science*, and *Global Environmental Change*. He has recently contributed to the *Routledge Handbook of Media, Conflict and Security*, focusing on environmental security.

Libby Lester is Professor of Journalism, Media and Communications, and Associate Dean, Research for the College of Arts, Law and Education at the University of Tasmania. She has authored, co-authored, and co-edited six books, including *Leadership and the Construction of Environmental Concerns* (forthcoming, Palgrave Macmillan), *Environmental Pollution and the Media: Political Discourses of Risk and Responsibility in Australia, China and Japan* (Routledge 2017), *Media and Environment: Conflict, Politics and the News* (Polity 2010; Arabic ed 2013), and *Transnational Protests and the Media* (co-edited with Simon Cottle).

Justin Lewis is Professor of Communication at Cardiff School of Journalism, Media and Culture. He has written widely about media, culture, and politics. His books, since 2000, include *Constructing Public Opinion* (2001), *Citizens or Consumers: What the Media Tell Us About Political Participation* (2005), *Shoot First and Ask Questions Later: Media Coverage of the War in Iraq* (2006), *Climate Change and the Media* (2009), and *The World of 24 Hour News* (2010). His latest book is *Beyond Consumer Capitalism: Media and the Limits to Imagination* (2013). He has also written books on media audiences, cultural policy, and media and race.

Eloisa Beling Loose is a journalist, lecturer and researcher in the field of journalism and environment in Brazil. She is currently a lecturer at the

Department of Communication of UFPR and a member of a research group on Environmental Journalism (UFRGS/CNPq). Her latest research is about the relations between communication, risk perception, and the governance of climate change, focusing on the actors involved in the production and reception of local news in Curitiba.

Paolo Magagnoli is a modern and contemporary art historian who specialises in the history of photography and the moving image. He is currently a Lecturer in contemporary art at the University of Queensland. He has worked as a curatorial assistant for the Department of Photographs of The Metropolitan Museum of Art and The American Federation of Arts in New York. His essays have been published in refereed academic journals such as *The Oxford Art Journal*, *Third Text*, *Philosophy of Photography*, *Transnational Cinemas*, and *Academic Quarter*. He is also a critic, writing for international art presses and magazines such as *Frieze*, *Art in America*, *Afterall*, and *Camera Austria*. His latest book is *Documents of Utopia* (Columbia University Press, 2015).

Alana Mann joined the University of Sydney in 2007 after a professional career in the media and non-profit sectors. Her teaching and research focus on how ordinary citizens get voice in policy debates regarding wicked problems such as food security and climate change. Her book *Global Activism in Food Politics: Power Shift* was published in 2014. Currently, Alana is involved in cross-disciplinary research projects concerning food systems with colleagues in the Sydney Environment Institute (SEI) and the Charles Perkins Centre. Her international collaborations include a comparative study of 'land-grabbing' with researchers in Brazil and South East Asia.

Richard Maxwell is a political economist of media. His research begins at the intersection of politics and economics to analyse the global media, their social and cultural impact, and the policies that regulate their reach and operations. He has published widely on a range of topics, from television in Spain's democratic transition to Hollywood's international dominance, from media politics in the post 9–11 era to how big political economic forces work in the mundane routines of daily life and culture. His current work on the environmental impact of media focuses on the environmental harms caused by media, information technologies, and electronics. His recent publications include *Greening the Media* (with Toby Miller), *Routledge Companion to Labor and Media*, and *Media and the Ecological Crisis* (co-edited with Jon Raundalen and Nina Lager Vestberg).

Lyn McGaurr is a Research Associate in the School of Social Sciences at the University of Tasmania, where she completed a PhD in Journalism, Media and Communications. She is the author of *Environmental Communication and Travel Journalism* (Routledge) and the co-author, with Bruce Tranter and Libby Lester, of *Leadership and the Construction of Environmental Concern* (Palgrave Macmillan, forthcoming). Her research interests include climate change communication, and the way environmental movements in Australia and Canada frame tourism.

Toby Miller is Distinguished Professor of the Graduate Division, University of California, Riverside; Sir Walter Murdoch Professor of Cultural Policy Studies, Murdoch University; Profesor Invitado, Escuela de Comunicación Social, Universidad del Norte; Professor of Journalism, Media and Cultural Studies, Cardiff University/Prifysgol Caerdydd; and Director of the Institute for Media and Creative Industries, Loughborough University London. The author and editor of over forty books, his work has been translated into Spanish, Chinese, Portuguese, Japanese, Turkish, German, Italian, Farsi, and Swedish. His most recent volumes are *Greenwashing Culture* (2018), *Greenwashing Sport* (2018), *The Routledge Companion to Global Cultural Policy* (edited with Victoria Durrer and Dave O'Brien, 2018), *Global Media Studies* (with Marwan M. Kraidy, 2016), *The Sage Companion to Television Studies* (edited with Manuel Alvarado, Milly Buonanno, and Herman Gray, 2015), *The Routledge Companion to Global Popular Culture* (edited, 2015), *Greening the Media* (with Richard Maxwell, 2012), and *Blow Up the Humanities* (2012).

Dr James Painter is a research associate at the Reuters Institute for the Study of Journalism (RISJ), Department of Politics, Oxford University. Before joining the RISJ, he worked for the BBC as a correspondent, and in senior management positions. He is the author of several books and academic articles on climate change and the media, including *Summoned by Science: Reporting Climate Change at Copenhagen and Beyond* (2010); *Poles Apart: The International Reporting of Climate Scepticism* (2011); *Climate Change in the Media: Reporting Risk and Uncertainty* (2013); and *Something Old, Something New: Digital Media and the Coverage of Climate Change* (2016).

Dr Mike S. Schäfer is professor of science, crisis, and risk communication at the Department of Mass Communication and Media research at the University of Zurich. He is also Director of the university's Center for Higher Education and Science Studies (CHESS), associate editor of the *Oxford Encyclopedia of Climate Change Communication* and serves on the editorial boards of *Public Understanding of Science, Environmental Communication,* and *JCOM—Journal of Science Communication.* He has published widely on climate change communication, most recently *Something Old, Something New* (Reuters Institute for the Study of Journalism, 2017, with James Painter et al.) and *Framing Analyses in Climate Change.*

Lorraine Whitmarsh is an environmental psychologist, specialising in perceptions and behaviour in relation to climate change, energy, and transport, based in the School of Psychology at Cardiff University. She is also theme leader for the Tyndall Centre for Climate Change Research. She regularly advises governmental and other organisations on climate change communications and behaviour change. Her research projects have applied both qualitative and quantitative methods to examine climate change scepticism, responses to flooding and sea-level rise, energy efficiency behaviours, travel habits, perceptions of smart technologies, and related issues.

Dr Terry Woronov is Senior Lecturer in Anthropology at the University of Sydney. She has lived, worked, and studied in China, Hong Kong, and Taiwan for many years. Dr Woronov holds a Ph.D. in Anthropology from the University of Chicago, an M.Ed. from Harvard, and was a post-doctoral research fellow at the Institute for East Asian Studies at the University of California-Berkeley. She is the author of *Class Work: Vocational Schools and China's Urban Youth* (2015, Stanford University Press).

GLOBAL CRISES AND THE MEDIA

Simon Cottle, *General Editor*

From climate change to the war on terror, financial meltdowns to forced migrations, pandemics to world poverty, and humanitarian disasters to the denial of human rights, these and other crises represent the dark side of our globalized planet. They are endemic to the contemporary global world and so too are they highly dependent on the world's media.

Each of the specially commissioned books in the *Global Crises and the Media* series examines the media's role, representation, and responsibility in covering major global crises. They show how the media can enter into their constitution, enacting them on the public stage and thereby helping to shape their future trajectory around the world. Each book provides a sophisticated and empirically engaged understanding of the topic in order to invigorate the wider academic study and public debate about the most pressing and historically unprecedented global crises of our time.

For further information about the series and submitting manuscripts, please contact:

Dr. Simon Cottle
Cardiff School of Journalism
Cardiff University, Room 1.28
The Bute Building, King Edward VII Ave.
Cardiff CF10 3NB
United Kingdom
CottleS@cardiff.ac.uk

To order other books in this series, please contact our Customer Service Department at:

(800) 770-LANG (within the U.S.)
(212) 647-7706 (outside the U.S.)
(212) 647-7707 FAX

Or browse online by series at:

www.peterlang.com

www.ingramcontent.com/pod-product-compliance
Ingram Content Group UK Ltd.
Pitfield, Milton Keynes, MK11 3LW, UK
UKHW020159290926
15471UKWH00005B/967

9 781433 151330